OXFORD–WARBURG STUDIES

General Editors

DENYS HAY *and* J. B. TRAPP

OXFORD–WARBURG STUDIES

The Conflict between Paganism and Christianity in the Fourth Century
Essays edited by A. D. MOMIGLIANO. 1963

Jacopo della Quercia's Fonte Gaia
By ANNE COFFIN HANSON. 1965

The Sistine Chapel before Michelangelo
By L. D. ETTLINGER. 1965

The Government of Florence under the Medici 1434–1494
By NICOLAI RUBINSTEIN. 1966

Spectacle, Pageantry, and Early Tudor Policy
By SYDNEY ANGLO. 1969

Giotto and the Orators: Humanist Observers of Painting in Italy and the Discovery of Pictorial Composition
By MICHAEL BAXANDALL. 1971

Raymond Lull and Lullism in Fourteenth-century France
By J. N. HILLGARTH. 1972

The Manifold in Perception: Theories of Art from Kant to Hildebrand
By MICHAEL PODRO. 1972

The Figurae of Joachim of Fiore
By MARJORIE REEVES and BEATRICE HIRSCH-REICH. 1972

Renovation and Counter-Reformation: Vasari and Duke Cosimo in Sta Maria Novella and Sta Croce
By MARCIA B. HALL. 1979

The Fall of the Florentine Republic 1512–30
By J. N. STEPHENS. 1983

Joseph Scaliger: A Study in the History of Classical Scholarship, I. Textual Criticism and Exegesis
By ANTHONY GRAFTON. 1983

LATIN POETRY AND THE CLASSICAL TRADITION

Essays in Medieval and Renaissance Literature

EDITED BY
PETER GODMAN
AND
OSWYN MURRAY

CLARENDON PRESS · OXFORD
1990

Oxford University Press, Walton Street, Oxford OX2 6DP
Oxford New York Toronto
Delhi Bombay Calcutta Madras Karachi
Petaling Jaya Singapore Hong Kong Tokyo
Nairobi Dar es Salaam Cape Town
Melbourne Auckland
and associated companies in
Berlin Ibadan

Oxford is a trade mark of Oxford University Press

Published in the United States
by Oxford University Press, New York

British Library Cataloguing in Publication Data
Latin poetry and the classical tradition: essays in medieval and Renaissance literature.—(Oxford–Warburg studies).
1. Poetry in Latin, 750–1350—Critical studies
I. Godman, Peter II. Murray, Oswyn III. Series.
871.0309
ISBN 0–19–920174–9

Library of Congress Cataloging in Publication Data
Latin poetry and the classical tradition: essays in medieval and Renaissance literature / edited by Peter Godman and Oswyn Murray.
p. cm.—(Oxford–Warburg studies)
"This book has grown out of a conference, held at the Warburg Institute in December 1988"—Pref.
Includes index.
1. Latin poetry, Medieval and modern—History and criticism—Congresses. 2. Latin poetry, Medieval and modern—Classical influences—Congresses. 3. Renaissance—Congresses. 4. Classicism—Congresses. I. Godman, Peter. II. Murray, Oswyn. III. Series.
PA8050.L38 1990
871'0309—dc20 90–7727
ISBN 0–19–920174–9

Typeset by Hope Services (Abingdon) Ltd.
Printed and bound in
Great Britain by Biddles Ltd.
Guildford and King's Lynn

JOSEPHO TRAPP

harum litterarum fautori
humanissimo

Cuinam itaque potius convenire possunt, quàm Illi qui Ætate florens, & politissimo Ingenio, exquisitae in Civilibus Negotiis tractandis peritiae humaniorem Literaturam conjungit; Poetarum antiquorum veneres & impense amat, & perfectè intellegit?

J. TRAPP, *Praelectiones Poeticae* (Oxford, 1711), Epistola Nuncupatoria

Preface

THIS book has grown out of a conference held at the Warburg Institute in December, 1988. Its central theme is the character of the classical tradition in Latin poetry of the Middle Ages and the Renaissance. Through the different but complementary approaches of an international team of scholars, it aims to shed light on the role of the post-classical Latin poet as a creative interpreter of antiquity's heritage. No attempt has been made to impose a unified line on the contributors' treatment of their chosen topics, nor does this volume pretend to comprehensiveness. But enough is said, about a sufficiently wide range of periods and problems, to evoke the richness and diversity of the theme of *Latin Poetry and the Classical Tradition.*

The editors acknowledge with thanks the generous support of the Warburg Institute and of the British Academy, the friendly co-operation of the contributors, and the efficiency of the Press in ensuring swift publication. That this book exists at all is due chiefly to the energy and acumen of the Warburg's director. By dedicating it to him, in the year of his retirement, we wish not only to record, with gratitude and affection, a personal debt but also to signal the wider obligations owed to J. B. Trapp by the world of scholarship.

P.G. and O.M.

Oxford

January 1990

Contents

List of Contributors

A. C. DIONISOTTI is Lecturer in Classics, King's College, University of London.

PETER DRONKE is Professor of Medieval Latin, University of Cambridge.

MICHELE FEO is Professor of Medieval and Renaissance Latin, University of Florence.

PETER GODMAN is Professor of Medieval Latin, Tübingen University.

JOZEF IJSEWIJN is Professor of Humanistic Philology, Louvain University.

WALTHER LUDWIG is Professor of Classical Philology, Hamburg University.

OSWYN MURRAY is a Fellow of Balliol College, Oxford.

GIOVANNI ORLANDI is Professor of Medieval Latin, Milan University.

G. W. PIGMAN III is Professor of English and Comparative Literature, California Institute of Technology.

PAUL GERHARD SCHMIDT is Professor of Medieval Latin, University of Freiburg-im-Breisgau.

JAN M. ZIOLKOWSKI is Professor of Medieval Latin, Harvard University.

I The Idea of the Shepherd King from Cyrus to Charlemagne

OSWYN MURRAY

The old idea that the culture of the past explains the present belongs to a Darwinian view of history, which sees cultures as developing from origins according to deterministic principles. Michael Baxandall has recently protested against such an implication in relation to the concept of artistic influence:

> 'Influence' is a curse of art criticism primarily because of its wrong-headed grammatical prejudice about who is the agent and who the patient: it seems to reverse the active/passive relation which the historical actor experiences and the inferential beholder will wish to take into account. If one says that X influenced Y it does seem that one is saying that X did something to Y rather than that Y did something to X.
>
> (*Patterns of Intention* (New Haven–London, 1985), pp. 58 f.)

The same objection can be raised against the concept of 'tradition'. Tradition is often represented as moulding the perceptions, directing energies along already determined paths, and therefore as a historical force which both intensifies and limits artistic creativity. Yet it is the *use* of the past, not the copying of the past, which constitutes a genuine tradition. The 'classical tradition' in particular has been marked by this tendency, because of three inter-related phenomena: these are the importance of discontinuity as much as continuity in the development of the Latin culture of medieval and Renaissance Europe, the consequent emphasis on the idea of rebirth or renewal,[1] and the potentiality for creative misunderstanding implicit in this process. 'Classicism' has been only one aspect of the classical tradition, and it has been an aspect not without danger for the continuity of that tradition. For in fact the

[1] E. Panofsky, *Renaissance and Renascences in Western Art* (Stockholm, 1965).

classical tradition has been so important for western culture precisely because it has been so frequently remoulded, giving each separate age a sense of belonging to a tradition, combined with freedom of expression. The concept of freedom is also intrinsic to the relationship between classical tradition and Christianity; for in the past the classical tradition has usually been seen to represent a challenge to the prevailing ideology, just as it does to modern materialism. A more continuous, more dominant, less subversive tradition would not have served the liberating function of a counter-culture, and would have created a concern with norms rather than with their manipulation. The artist must use the past to create the future, and that use must involve an element of misuse. The conception of tradition as a creative force, a liberating factor in the culture of Latin Europe, enabling new forms to emerge, is basic to the meaning of the classical tradition in relation to medieval and Renaissance Latin literature.

The example I have chosen seeks to illustrate three points, first of discontinuity, second of the fusion of the different influences of classicism and Christianity, and third of the creative use of the past. It also concerns the relationship between myth and literary images. In fact, it seems to me that myth and poetic imagery operate in much the same way: they offer the impression of continuity through the process of historical change. And since the human situation and the human psyche impose the phenomenon of recurrence on the diversity of history, myth comes to possess a resonance, a potentiality for the ordering of experience across ages and across cultures;[2] yet it is not immune from the same forces which control the imagery of poetry: it too is subject to the human will, the creative use (or misuse) of the inherited imagination.

It is usual to conceive of the idea of the shepherd king as involving a constant image, which can be traced from the origins of civilization in the ancient Near East through classical and Christian forms into the Middle Ages and the Renaissance, and beyond to the age of Mozart's *Re pastore* and the *fête champêtre*, as far perhaps as Picasso's late studies of Manet's *Déjeuner sur l'herbe*. But this synoptic vision is misleading, as I hope to show. If I stop with the Carolingian age, it is partly because the later pastoral tradition is more continuous, and has been more studied,[3] but

[2] For the artist's viewpoint see David Jones, 'Art and Sacrament', in *Epoch and Artist* (London, 1959), 143–79.

[3] Most recently by Annabel Patterson, *Pastoral Ideology: Virgil to Valéry* (Oxford, 1988); but see the review of R. Jenkyns, *Essays in Criticism*, 39 (1989), 65–72. For the period

perhaps as much because the theoretical points I wish to make will already have been sufficiently exemplified.

I

The positive image of the king as shepherd or herdsman of his people is not a universal one; it is, rather, specific to those cultures which derive more or less directly from the ancient Near East. There it appears already as the commonest royal 'descriptive title' as early as the Sumerian King List;[4] and in those rare moments when the Babylonian and Assyrian inscriptions dwell on the good government of their kings, rather than their ability to strike terror into their enemies, they are described as 'the true shepherd',[5] or one 'whose shepherding the gods made as agreeable to the people of Assyria as is the scent of the plant of life'.[6] As a royal designation, it can also be predicated of the gods. In Egypt, by contrast, the description is initially used of gods, but from the early Middle Kingdom onwards the king too is described as a shepherd: 'He appointed me shepherd of this land, knowing him who would herd it for him', says Sesostris I.[7]

These are isolated images, and we do not know how far they may have wider implications within their cultural context; perhaps, given the dominant agricultural forms of the two river-valley civilizations, they may have remained relatively undeveloped, or be vestiges of some earlier nomadic existence. But such pastoral imagery developed into a more complex ideology precisely in those pastoral cultures which came into contact with Mesopotamia and Egypt.

The foundation myth of the Persian Empire was fashioned out of the traditional myth of the exposure of the king-child, which we know first from the legend of the exposure and upbringing of Sargon of Accad under the protection of the goddess Ishtar. In the Persian account the setting is transposed from the river banks and gardens of Mesopotamia to the nomadic uplands. Cyrus is

covered in the later part of this paper, cf. also Helen Cooper, *Pastoral*: *Medieval into Renaissance* (Ipswich, 1977), esp. ch. i.

[4] T. Jacobsen, *The Sumerian King List* (Chicago, 1939), cols. i. 15; ii. 16; iii. 12; see p. 142.

[5] J. B. Pritchard, *Ancient Near Eastern Texts relating to the Old Testament* (Princeton, 1955), 289, col. i (Esarhaddon, 680–669 BC); cf. col. ii: 'like lambs they gamboled and recognized me as their lord by praying to me'.

[6] ib. 281 (Adad-Nirari III, 810–783 BC).

[7] M. Lichtheim, *Ancient Egyptian Literature*, i (Berkeley, 1973), 116.

exposed as a baby, protected (or suckled) by the dog sacred to Mithras, and grows up amongst the herdsmen, before revealing himself as chosen by the god to be king. The story is told in a number of variants by Greek sources, Herodotus, Ctesias, and others; that it was the official Persian dynastic myth is shown by the ritual of the Persian coronation: the king had to go to Pasargadae, the original capital, where the Magi supervised a ceremony in which the king put on the original clothes of Cyrus as a herdsman and ate a herdsman's meal of figs, terebinth root, and sour milk.[8] The myth and its attendant ritual serve two purposes: the first is common to many royal exposure myths, which seek to explain the rise of a new power by emphasis on the new hero and his special relationship with the divine. The second is perhaps more characteristic of nomadic cultures, with their uneasy relationship to imperial wealth: it served to remind the dynasty of those origins which Cyrus symbolized; so Herodotus ends his great *History* with a conversation in which Cyrus warns the Persians to remain in their barren hills, and not succumb to the temptations of luxury.

This particular myth has had its influence, on later Sassanian royal mythology, and (through its Greek versions) on various western foundation myths, such as the legend of Romulus.[9] It helps too to explain that most famous and most puzzling appearance of shepherds, in Saint Luke's version of the Nativity; for his narrative contains many elements related to the nexus of myths concerning the exposure of the king-child: as Gressmann argued long ago, it is the shepherds who are in a certain sense a primary structural element in the story, and the Holy Family which is secondary.[10]

It is of course the Old Testament which provides the most developed ideology of the shepherd-king. Like other Near Eastern monarchic motifs, this was adapted to express the care of Jehovah for his chosen people. He became, especially in the Psalms and the Prophets, 'the Shepherd of Israel' (Genesis 49: 24), who leads his people and finds them pasture (Jeremiah 50: 19); the most complete and most moving expression of that image is the Psalm which we all know by heart (Psalm 23). But Jehovah can also refer

[8] Plutarch, *Artaxerxes* 3; A. Alföldi, 'Königsweihe und Männerbund bei den Achämeniden', *Schweizerisches Archiv für Volkskunde*, 47 (1951), 11–16; G. Binder, *Die Aussetzung des Königskindes* (Meisenheim, 1964); O. Murray, *Classical Review* 17 (1967), 329–32.

[9] Binder, op. cit.

[10] H. Gressmann, *Das Weihnachts-Ev auf Ursprung und Geschichte untersucht* (Göttingen, 1914), 16–28.

to his chosen king as a 'shepherd of the people' for good (Jeremiah 3: 15) or ill (2: 8; 10: 21; 11: 10): 'My people hath been lost sheep: their shepherds have caused them to go astray' (Jeremiah 50: 6, cf. 17; Zechariah 10: 13); and foreign kings too can be treated in the same way: Cyrus is 'my shepherd', while Jeremiah prophesies 'Howl ye shepherds, and cry' at the wrath of God (25: 34 ff).

The image of the king as shepherd is especially developed by the prophets in relation to Messianic hopes, of which perhaps the finest expression is Ezekiel 34. 'My servant David shall become king over them and they shall have one shepherd' (37: 24): what relation such prophecies have to the David of the historical books, who 'kept his father's sheep' (1 Samuel 17: 34, etc.), is a question for the experts. Here I only wish to emphasize that the biblical image of the shepherd-king, whether it refers to God or men, is by far the most coherent that survives from the ancient world, and occupies a place that appears to be far more central in the conception of kingship than for any earlier civilization. We should not obscure the individuality of the Israelite idea of kingship in our search for its origins. It is this Jewish tradition which Philo sums up in his discussion of pastoral metaphor in *De agricultura* 50:[11]

So good a thing is shepherding that it is rightly ascribed, not only to kings and wise men and souls of perfect purity, but also to Almighty God. The authority for this is not a chance one, but a prophet whom it is right to trust, he who wrote the Psalms. For he says this, 'The Lord is my shepherd, and I shall not want'.

The New Testament conception of the Good Shepherd in John 10 is a development from this fusion of divine and human kingship in the Messianic ideal of the Old Testament;[12] other references in the gospels often build explicitly on earlier prophecies like that in Mark 14: 27 and Matthew 26: 31 concerning the bad shepherd of Zechariah 13. And the developing conception in the early Church of the Christian community as Christ's flock, or of its leaders as shepherds seems a natural extension of the earlier biblical tradition.[13]

[11] Cf. J. Quasten, 'Der Gute Hirte in hellenistischer und frühgriechischer Logostheologie', in *Heilige Überlieferung: Festschrift I. Herwegen*, ed. O. Casel (Münster, 1938), 51–8.

[12] See R. Bultmann, *The Gospel of John: A Commentary* (Eng. trans., Oxford, 1971), 358–91; but his claim of gnostic origins (pp. 367–71) is unnecessary.

[13] Cf. P. Beskow, *Rex Gloriae: The Kingship of Christ in the Early Church* (Stockholm, 1962), 88 f., 198–200, with bibliography. For the early Christian artistic tradition see J. Kollwitz, 'Christusbild', *Reallexikon für Antike und Christentum* iii (1957), 5 f., 11 f., 23 f., under 'Hirte'.

II

In contrast it must be admitted that the image of the herdsman is by no means as important in the Greek tradition as modern writers, influenced by their Christian education and by the desire to find relations between Christianity and classical culture, have suggested. It is true that the image existed in the Homeric world, in the formulaic phrase of Homeric epic referring to kings, *poimena laōn*, found in accusative and dative as a line ending; and this phrase, itself perhaps a reflection of influence from the East, ensured that the comparison of king and shepherd was part of a common stock of literary motifs. But little is made of this epithet in the long tradition of Homeric exegesis, as far as Bishop Eustathius of Thessalonica in the twelfth century;[14] and it was not particularly emphasized in the kingship literature which developed from the fourth century BC onwards. It took its place alongside a stock of other metaphors for royal power, such as the father, the doctor, the pilot of a ship, or the sun among the stars.[15]

The most developed uses of the image of the king as shepherd are in fact to be found in Xenophon and Plato; and they concentrate more on the benefit to be drawn from being a herdsman than on the mutual relationship between shepherd and flock: thus Xenophon contrasts the ease of controlling docile animals with the problems of governing a people (*Cyropaedia* 1. 1), and his hero Cyrus compares the task of king with that of herdsman, in that both combine care for animals with use of them (8. 1. 14). In the famous discussion in *Republic* I, Thrasymachus ridicules the idea that shepherds work for the benefit of their flock rather than for their own profit; and Socrates' counter-claim is scarcely convincing—that the true art of the shepherd involves only that part of his task which concerns the care of his flock (*Republic* 343 ff.). Again in the *Statesman* the view that kingship and herdsmanship belong to the same logical division of 'ruling' immediately leads to the rueful statement that the herdsman has a freer hand in the selective breeding and culling of his herd (267 D).

The conception of the herdsman in classical political thought is

[14] See *Scholia Graeca in Homeri Iliadem*, ed. H. Erbse, i (Berlin, 1969), on *Iliad* 2.85b; *Eustathii Commentarii ad Homeri Iliadem pertinentes*, ed. M. van der Valk, i (Leiden, 1971), on 2.243 (= p. 212. 31 ff.). Eustathius discusses the differences between shepherd, goatherd, and cowherd in relation to kingship, in a passage which seems to go back to Philo.

[15] G. Barner, *Comparantur inter se Graeci de regentium hominum virtutibus auctores* (Diss. Marburg, 1889).

a thoroughly practical one, concerned with control and profit, and involving no positive ideology and no complex imagery. This is not perhaps surprising; for, of all animals, sheep possess a docility which tends to be more highly prized in a monarchy or a theocracy than in a democratic city-state: so Aristophanes presents a nightmare of demagogic rule, in which the Athenian assembly appears as a flock of sheep, being harangued by a voracious whale with the voice of an enraged sow (*Wasps* 31–6).

III

In classical antiquity it is the *Eclogues* of Virgil which bring together most successfully the world of herdsmen with that of kings; though it is important to remember that Augustus does not himself appear as a shepherd in Arcadia, but is rather conceived of as some external tutelary deity or guarantor of the pastoral peace—like an absentee landlord. The modern reading of the *Eclogues* tends to play down their political content in the interests of a true 'Virgilian' interpretation.[16] It treats as an unfortunate aberration the view of the Virgilian commentators of late antiquity, Donatus and Servius, that the poems are in large part allegories, which have a double *intentio*. According to these commentators, in the first place, 'intentio . . . in imitatione Theocriti poetae constituitur' ('the aim consists in imitation of the poet Theocritus'); in the second place 'est intentio etiam in laude Caesaris et principum ceterorum, per quos in sedes suos atque agros rediit' ('the aim is also praise of Caesar and the other nobles, through whom he returned to his home and fields').[17] Virgil differs in this respect from Theocritus, who is 'ubique simplex'. Yet in the history of the reception of Virgil, it is the political interpretation of the scholiasts which has been dominant until the Romantic period, and which has strongly influenced the range of uses to which pastoral has been put.[18]

The theory of Donatus and Servius is expounded both in Donatus' introduction to the *Eclogues*,[19] and through the elaborate

[16] See F. Leo, 'Vergils erste und neunte Ecloge', *Ausgewählte Kleine Schriften*, ii (Rome, 1960), 11–28; G. Williams, *Tradition and Originality in Roman Poetry* (Oxford, 1968), 303–29.

[17] *Vita Donati* 64; cf. Servius, p. 2 Thilo.

[18] Patterson (n. 3 above); despite the criticisms of Jenkyns, Patterson has some interesting things to say, especially on Servius in ch. i.

[19] This appears at the end of the Donatan Life of Virgil, and is most easily accessible in the editions of E. Diehl, *Die Vitae Vergilianae und ihre antiken Quellen* (Bonn, 1911), 22–6, or J. Brummer, *Vitae Vergilianae* (Leipzig, 1912), 11–19.

and detailed commentary of Servius, which often rejects the more extreme attempts to identify allegory, in the interests of maintaining the double *intentio* of the poet;[20] for

illud tenendum esse praedicimus, in bucolicis Vergilii neque nusquam neque ubique aliquid figurate dici, hoc est per allegoriam, vix enim propter laudem Caesaris et amissos agros haec Vergilio conceduntur, cum Theocritus simpliciter conscripserit, quem hoc noster conatur imitari.

we declare this principle must be adhered to, that figurative language (that is allegory) is used in Virgil's bucolics neither nowhere nor everywhere; for concern with praise of Caesar and his lost estate is only partially permitted to Virgil, since Theocritus wrote with a single aim, and our author is seeking to imitate him.

The 'character' of bucolic is 'humilis' in the threefold category, 'humilis, medius, grandiloquus', and could not therefore be entirely concerned with praise, for *encomium* naturally belongs to the third stylistic level:[21] Virgil himself recognizes this in the opening of his fourth *Eclogue*—'Sicelides musae, paulo maiora canamus!'[22]

The Donatan reading of the *Eclogues* is not of course an invention of late antiquity. It appears as standard educational orthodoxy in the late first century, when Quintilian quotes lines of the ninth *Eclogue* to illustrate the notion of 'allegoria sine translatione', holding that in the description of Menalcas' estate we should understand 'Vergilius' for 'Menalcas'.[23] In the moderate form in which it is couched by the commentators, 'neque nusquam neque ubique', it attests a long tradition of such interpretation, with a variety of views on the extent of the panegyrical element; for Donatus and Servius are as much concerned with rejecting extreme versions of the theory as with establishing their own positions, and Servius seems anxious to play down even the claims of Donatus. This reading is also significantly different in emphasis from that which modern scholarship presumes to criticize. For it is primarily concerned with the place of *laudes* in the poems, not with events in the life of the poet: this latter view is a modern post-Romantic biographical interpretation of what was in origin a theory about literary genre and decorum.

[20] *Vita Donati* 66; compare Servius: 'non tamen ubique sed tantum ubi exigit ratio' (p. 4. 22 Thilo).

[21] Servius, p. 2. 1 Thilo; or in Donatus' formulation', 'tenuis, moderatus, validus' (with their Greek originals), *Vita Donati* 58–9.

[22] *Vita Donati* 65.

[23] Quintilian 8. 6. 44–7: 'non pastor Menalcas, sed Vergilius est intellegendus.'

As such, the interpretation goes back at least to the generation after Virgil, if not indeed to the original reception of the work. For it served as the presupposition of the first remarkable revival of pastoral known to us, in the Neronian age.[24] This was a court poetry, practised by men like the future emperor Nerva, 'the Tibullus of our time' according to Nero;[25] and the two poets whose work we possess show how essential panegyric was believed to be to the Virgilian pastoral. Calpurnius uses the genre to praise Nero with greater directness than Virgil, but in accordance with the prescription 'neque nusquam neque ubique'. His collection of poems is intentionally framed by praise of the emperor: in *Eclogue* 1 his shepherds step out of the pastoral scene almost self-consciously, as they find a divine prophecy of the new Golden Age carved on a rustic tree; in *Eclogue* 7 news of a visit to the great city of Rome and its theatre culminates in rustic admiration of the divine emperor. *Eclogue* 4 offers a poetic contest between two shepherds singing in praise of the emperor. These poems shall be carried by another to the emperor's court, and win the poet favour, as 'he who brought Tityrus of the sweet song from the woods to the queen of cities, and showed him the divine powers, and said, "scorning the sheepfold, Tityrus, we will sing first of the countryside, but after of the arms of war"' (4. 160–3).

Calpurnius juxtaposes the various elements; the two bucolic fragments known as the *Carmina Einsiedlensia* seek to fuse them in a form which suggests that true allegory was the poet's intention. The first fragment offers the beginning of a poetic contest in which the inspiration derived from Apollo succumbs to that derived from the emperor: 'mine is the prize because my heart bids me speak Caesar's praise: the palm is ever due to such a task' (15). The second fragment has caused historians great problems, because it takes too seriously the Virgilian precedent, that something needs to be wrong in Arcadia, so that the god who puts it right can win praise. 'Quid tacitus Mystes?—curae mea gaudia turbant' (1). But what can possibly be wrong in an Arcadia at the service of panegyric? Only satiety apparently, and the poem turns to a description of the Golden Age that seeks to outdo Virgil's fourth Eclogue. We never come back to the cares which oppress our too thoughtful shepherd, doubtless because any attempt to do so would effectively undermine the entire project. The fact is that a

[24] There is a useful edition of these poets by D. Korzeniewski, *Hirtengedichte aus neronischer Zeit* (Darmstadt, 1971).

[25] Martial 8. 70; 9. 26. 1; Pliny, *Epistles* 5. 3. 5.

too literal attempt to copy the Virgilian relationship between misfortune and thanks for benefits received has led the author into a trap: there cannot be anything wrong in the pastoral vision of the Neronian Age. Calpurnius does better in following Seneca's *de clementia*, and praising the new age because it is not born of civil war and is therefore free from the cares which oppressed Virgil's Arcadia.[26]

IV

There is little political use of the image of the shepherd in late antiquity, despite (or perhaps because of) the Christianization of the empire.[27] In part this must reflect the continuing dominance of classical models which laid small emphasis on the king as shepherd; in part perhaps it reflects the fact that the Christian image was firmly related to the figure of Christ and the conception of leadership within the Church: its religious connotations were too strong to allow its use in secular contexts.

The particular form which the Carolingian ideology of kingship took is therefore new, even if we may regard it as the consequence of a fusion of two traditions. This ideology has been much studied from both the political and the cultural points of view. The comparison of Charlemagne with the biblical David has Byzantine forerunners, and is expressed as early as 775; but the regular use of 'David' in reference to him belongs to literary texts from about 794 onwards. Historians have discussed the points of contact, the concept of the king as the Lord's anointed, as prophet of his people, as victorious warrior, and as repository of wisdom.[28] But historians are always a little uncomfortable with poetry, and have tended to play down one aspect of Alcuin's famous comparison of Charlemagne and David (*MGH Epist.* iv, no. 41, p. 84):

Ita et David olim praecedentis populi rex a Deo electus et Deo dilectus et egregius psalmista Israheli victrici gladio undique gentes subiciens, legisque Dei eximius praedicator in populo extitit.

[26] Calpurnius Siculus 1. 33–88; cf. Seneca, *De clementia* 1. 9–11; *Apocolocyntosis* 4. 1.

[27] Passing references in the imperial panegyrists of the Greek East reflect the influence of Dio of Prusa's orations on kingship: see e.g. Dio, *Orationes* 1. 28; Themistius, *Orationes* 1. 10a; Synesius, *p.B.* 6a–b.

[28] For bibliography see P. Godman, *Poets and Emperors: Frankish Politics and Carolingian Poetry* (Oxford, 1987), 65 nn. 143–4.

So too David, once the king of a former people, chosen by God and beloved of God and famed psalmist of Israel, subduing the nations everywhere with his victorious sword, stood forth as noble preacher of the law of God to his people.

'Egregius psalmista': as Peter Godman has recently reminded us, 'harps mattered—symbolically, artistically, and therefore politically'.[29] The image of Charlemagne as poet is closely bound up with his name of David: 'David amat vates, vatorum est gloria David' (Angilbert 2. 3, *MGH Poetae*, i. 360–3).

The court poetry of which he is the prince differs from that earlier Merovingian poetry created by Venantius Fortunatus. Venantius is the latest product of a Latin tradition of poetic panegyric which can be traced back to Claudian; but, despite the knowledge of Venantius and the respect paid to him in the Carolingian age, this court poetry is free of the last constraints of rhetorical panegyric, and is based on poetic models. It derives primarily from a reading of Virgil, and especially of the *Eclogues*.

The *Eclogues* of the Carolingian poets are not of course our *Eclogues*: they are the *Eclogues* as expounded by the tradition of Servius. As Curtius said, 'from the first century of the Empire to the time of Goethe, all study of Latin literature began with the first eclogue'.[30] On the first line of the first *Eclogue*, 'Tityre tu patulae recubans sub tegmine fagi', the edition of the scholia by Pierre Daniel offers the gloss 'a lite, O Virgili, sub protectione Augusti securus quiescis' ('Virgil, you lie safe from the dispute under the protection of Augustus'), and a little later 'in umbra allegorice sub tutela Imp. Aug.' ('in the shade—allegorically, under the protection of the emperor Augustus').[31] This interpretation is surely the basis of Carolingian uses of the phrase: the deeds of earlier kings 'lateant Karoli sub tegmine regis' (Peter of Pisa 41. 4, *MGH Poetae*, i. 74); this page contains praises 'which I Theodulf sang under the name of the king', ('quas ego Teudulfus cecini sub nomine regis', Theodulf 23. 39, *MGH Poetae*, i. 481).

[29] Loc. cit. On Charlemagne and David see esp. J. Fleckenstein, *Die Bildungsreform Karl des Grossen* (Bugge-Ruhr, 1953), 37–9, 57 ff.

[30] E. R. Curtius, *European Literature and the Latin Middle Ages* (Eng. trans., London, 1953), 190.

[31] In which manuscript Daniel found these and other clearly ancient comments on the *Eclogues* is unclear; they appear in his famous Paris edition of 1600 (p. 2. c); but Thilo did not find them in the MSS used by him, and relegated them to the *apparatus criticus* in his edition of Servius. On Daniel see H. Hagen, 'Der Jurist und Philolog Peter Daniel', *Festrede, Universität Bern* (1873).

It is the Servian reading of the *Eclogues* that lies behind the ubiquitous use of pastoral imagery in Carolingian court poetry. Angilbert can use the bucolic refrain just quoted, 'David amat vates, vatorum est gloria David'; Alcuin begins a poem to Pippin, Charlemagne's son, 'Pergite Pierides' (Alcuin 14, *MGH Poetae*, i.237), and paints a picture of Castalian springs and fields full of flowers ready to be plucked to garland the young hero. So Charlemagne is addressed, 'Rex Carole felix et notus in omnibus arvis' (Wigbod 8b. 2, *MGH Poetae*, i.96). The characteristic words in classical literature for the pastoral style, 'humilis', 'horridus', 'rudis', 'vilis', 'raucisonus', 'tenuis', 'gracilis', 'silvestris', but above all 'rusticus', serve to define a newly sophisticated style of poetry, which professes mock humility in the presence of exalted themes.[32]

The court circle is conceived of as a bucolic fraternity, but it is a fraternity which unites the worlds of Virgil and the Bible: ancient poets mingle with biblical figures in a new pastoral landscape. The central and unifying figure in this dual vision of pastoral is Charlemagne himself as king David the Psalmist, who combines the attributes of the shepherd-king with those of the shepherd-poet.

This poetry has recently been well discussed in its historical context by Peter Godman,[33] and I wish only to dwell on one of the poets he analyses, Moduin.[34] Moduin's two 'books' of *Eclogues* sum up the place of pastoral in Carolingian panegyric, and reveal how self-conscious and professional it had become by the end of Charlemagne's reign. The book is dedicated to Charlemagne, who especially will be the judge: here your fame is revealed 'carmine velato' (*Prol.* 6): for like the Virgilian eclogue of the Servian tradition, it is a 'carmen velatum' whose meaning is revealed 'allegorice'. The first *Eclogue*, a poetic dialogue modelled on Virgil's first *Eclogue* is in fact a defence of the new *Poesie*. The *puer* wishes to obtain wealth and honour by approaching the court with his pastorals; the *senex*, who is portrayed as both poet and retired warrior, seeks to dissuade him. The arguments of the *senex* seem to oscillate between attack on pastoral ('rustica raucisonae meditaris carmina Musae', 29) as if there were another better type of poetry—presumably his own—and attack on poetry in

[32] Cf. Curtius (n. 30 above), 83–5: 'Affected Modesty'.

[33] Op. cit. (n. 28 above), ch. iii.

[34] *MGH Poetae*, i. 385 ff.; cf. D. Korzeniewski, *Hirtengedichte aus spätrömischer und karolingischer Zeit* (Darmstadt, 1976).

general; for such reasons I would believe him to be no particular person, but an amalgam of those attitudes hostile to the new literature. That is why he is warrior as well as poet; and when he says, 'publica nulla canis, nulli tua carmina digna' (33), it is impossible to tell whether what he is condemning is all poetry, or merely pastoral poetry. Two of the arguments with which the 'puer' replies are especially effective: Charlemagne himself is often accustomed to play on the woodland reed ('ille solet calamo silvestri ludere saepe', 46); and both in antiquity and in the present generation many have found fame and wealth through their poetry, from Virgil to the older Carolingian court poets under their 'antique' names—Homerus (Angilbert), Flaccus (Alcuin), Theodulfus, Nardus (Einhard) (71–95). Moduin's poem is not so much a plea for entry into a world from which he as the *puer* is excluded, but rather a celebration of that world to which the poet has full right to belong, a professional world of honour and power. The *senex* represents the voices of envy outside that world, not of an insider trying to keep a newcomer out. And it is portrayed as a world in which pastoral and panegyric are synonymous.

The second of Moduin's *Eclogues* is a carefully worked panegyric in the pastoral manner, chiefly important for being modelled on Calpurnius' first *Eclogue*, thus revealing the surprising fact that the text of Calpurnius and Nemesianus was available in the Carolingian age, and was therefore one further support for the political reading of pastoral. The epilogue shows that the collection is a presentation volume for the king, which he is required to read to the end and correct: the image of the king as fellow shepherd-poet mingles with that of patron of the new poetry.

From this brief survey the idea of the shepherd-king may seem to emerge as neither archetypal myth nor cultural image, more a history of fragments and chance meetings, 'a thing of shreds and patches'. The study of motifs, whether artistic or literary, does indeed usually reveal more about discontinuities than continuities: it is important both in exploding the idea of a developing universe of culture, and in reminding us of the similarity between discovery and rediscovery. The power of images can even impose continuity on disparate phenomena: it makes not the slightest difference that in fact the shepherd of the Middle East and the Mediterranean does indeed lead his flock, while the shepherd of northern Europe has, at least since the end of antiquity, driven his, through fear not love, and with the help of dogs and foul language.

If there remains a general meaning in this intermittent image of royalty, it lies in the fusion of the different forms in which it appears at different times, expressing the king's relation to his people or to his god, his attitude to the past and the present. King and shepherd stand at opposite poles of the social system: they present the contrast between absolute power, and a life in the wilderness beyond the boundaries of society; yet in this opposition there is a common bond: both king and shepherd are to the ordinary man free spirits, outsiders, not bound by the normal social constraints. The composite picture therefore brings together savage origins and imperial splendour, wealth and the simple life, in an image especially suited to the charter myth or foundation legend.

The Carolingian age perhaps read that image more completely and more historically than any earlier or subsequent age, and expressed its meaning more successfully. For like the Persians before them, they too wished to unite the origins of their world in nomadic pastoralism with their role as rulers of a universal Christian Roman empire; and they intended the pastoral image as a new myth, to convey the place of the king in their world, both Frankish lord, and heir to the dual ancient heritage of David and Augustus. They knew their creation had roots (though they did not know the whole story), but that does not make it any less their creation. And it is because it is a new creation that it belongs to the continuing history of the classical tradition.

2 Classical Influences on Medieval Latin Views of Poetic Inspiration

JAN. M. ZIOLKOWSKI

Inspiration plays a part in the production of any poetic composition, no matter how drab or how gifted the poet.[1] Like many poets in many cultures, medieval Latin poets often explicitly requested inspiration and described the process of attaining it within their poems; and, like many poets, medieval Latin poets behaved in a somewhat paradoxical manner as they sought or studied creative inspiration: despite the intensely personal and private nature of the experience, they tended to be conditioned by conventions of their cultures. In particular, they responded to schemata inherited from classical Latin rhetoric and poetry as they identified places and times that lead to inspiration, as they chose the points in their writings at which to seek inspiration, as they singled out divine agents to invoke for inspiration, and as they described the physical and mental states likely to foster inspiration. At the same time, they modified classical conventions to bring them into conformity with Christian doctrine. It is also likely that sometimes they adapted classical schemata to suit conceptions of poetic inspiration that were current in their native cultures among poets who composed in vernacular languages.

In antiquity poetic inspiration had its own mythological landscape. According to Roman poets, the art of poetry was closely

I thank my colleagues John Bodel, Richard Thomas, and Calvert Watkins for discussing this topic with me, and Albert B. Lord for reading a draft. For helpful comments and references I am grateful to Penelope Murray. Bibliographical details of works cited in the notes are given below, pp. 35–8.

[1] For general remarks and bibliographic guidance see Benham; Chadwick, especially 1–14 and 41–57.

associated with the divinities and geography of two mountains. One was Parnassus, near Delphi. Of its two summits one was sacred to Apollo, the other to Dionysus. At the foot of a cliff was the Castalian spring, where the nymph Castalia had thrown herself to escape the pursuit of Apollo. This spring was sacred to Apollo and the Muses. The other mountain was Helicon, in Boeotia. Besides a grove where the Muses dwelt, it had two fountains sacred to the Muses, Aganippe and Hippocrene ('Horse-Fountain', Persius, *Satires*, prologue). The latter, so called because it was formed when the horse Pegasus struck the ground with his hoof (Ovid, *Metamorphoses* 5. 256–64), produced waters that were reputed to bring poetic inspiration. Long before the Middle Ages the Muses of Mount Helicon had become associated with mythological figures of other regions, most importantly with the Pierides and the Camenae but also with other nymphs.[2] None the less, the number of Muses had been standardized to nine: Calliope, Clio, Polyhymnia, Euterpe, Terpsichore, Erato, Melpomene, Thalia, and Urania. Although the specific functions assigned to these nine varied from one poet or mythographer to another, all were related to thought and artistic expression.

Because of geographical constraints, even in antiquity not every poet had the opportunity to ascend Mount Parnassus or Helicon to secure inspiration. Yet any poet could enter a grove or grotto and attempt to recreate the experience of an ascent, thereby propitiating the Muses. Furthermore, a poet could overcome physical distance from the Muses and achieve consecration in his art by dreaming of being transported to Helicon or Parnassus. Callimachus, Ennius, Virgil, Propertius, and Persius refer to the theme of the mountain-top epiphany.[3]

Although no Latin poets of the Middle Ages visited either Parnassus or Helicon, many mentioned the two mountains and their associated deities in their treatments of poetic inspiration. Furthermore, like the poets of Augustan Rome, many presented acts of composition as the results of encounters with the Muses in real or imagined groves and grottoes and with other divine figures within visions.[4]

Behind the elaborate apparatus of divinely inspired composition

[2] By the time of Isidore of Seville the terms *Musae* and *nymphae* were synonymous: *Etymologiae* 8. 11. 96.

[3] See especially Kambylis, as well as the notes on Horace, *Odes* 2. 19 (Nisbet and Hubbard, p. 315), and Persius, *Satires*, Prologue 2–3 (p. 64).

[4] Curtius, *European Literature*, 207–8, and Russell, 56.

in groves and grottoes is one uncontestable reality: the need of poets (and other authors!) for solitude in which to reflect and create. Because of this need Roman poets spoke of retiring from the cares of the city, which is noisy both night and day, to the glens and glades of the countryside. In the words of Horace (*Epistle* 2. 2. 78):[5]

> Scriptorum chorus omnis amat nemus et fugit urbem,
> rite cliens Bacchi somno gaudentis et umbra.
>
> The whole band of writers loves the grove and flees the city, duly devoted to Bacchus, who enjoys sleep and shade.

The preference of poets for glens and glades was canonized by the Romans in several discussions of literature. For instance, Tacitus declares in the *Dialogus de oratoribus* (9. 6):[6]

> Adice quod poetis, si modo dignum aliquid elaborare et efficere velint, relinquenda conversatio amicorum et iucunditas urbis, deserenda cetera officia, utque ipsi dicunt in nemora et lucos, id est in solitudinem recedendum est.
>
> And there is more. A poet, when he is minded laboriously to produce some creditable composition, has to turn his back on the society of friends and on all the charms of city life; abandoning every other function, he must retire into the solitude, as poets themselves say, of the woods and the groves.

Apparently in reaction to Tacitus, Pliny devotes one of his letters to urging a friend to take up hunting because 'the mere fact of being alone in the depths of the woods in the silence necessary for hunting is a positive stimulus to thought' (*Epistles* 1. 6 'iam undique silvae et solitudo ipsumque illud silentium quod venationi datur, magna cogitationis incitamenta sunt').[7] But as Quintilian points out with his ruthless practicality, the sylvan landscape that poets like to claim as the *locus* of their composition exists only in their imaginations. It is far removed from the reality of the techniques that most of them find productive (10. 3. 23):

> . . . atque liberum arbitris locum et quam altissimum silentium scribentibus maxime convenire nemo dubitaverit. Non tamen protinus audiendi qui

[5] Cf. Juvenal 7. 57 and Ovid, *Tristia* 1.1. 41.

[6] Cf. 12. 1.

[7] Contrast Pliny, *Epistles* 9. 10. 2 'itaque poemata quiescunt, quae tu inter nemora et lucos commodissime perfici putas' ('but peace reigns over the poems which you fancy are only too easy to finish in the woods and groves'). On the problematic interrelationship of Pliny's two letters and Tacitus' declaration, see Murgia, 179.

credunt aptissima in hoc nemora silvasque, quod illa caeli libertas locorumque amoenitas sublimem animum et beatiorem spiritum parent. Mihi certe iucundus hic magis quam studiorum hortator videtur esse secessus. Namque illa quae ipsa delectant necesse est avocent ab intentione operis destinati.

Everyone . . . will agree that the absence of company and deep silence are most conducive to writing, though Iwould not go so far as to concur in the opinion of those who think woods and groves the most suitable localities for the purpose, on the ground that the freedom of the sky and the charm of the surroundings produce sublimity of thought and wealth of inspiration. Personally I regard such an environment as a pleasant luxury rather than a stimulus to study. For whatever causes us delight, must necessarily distract us from the concentration due to our work.

What writers require is not trees and open air, but clear space (both physical and emotional) for concentration. This need for solitude also caused Quintilian to damn the custom of dictating to a secretary, since the presence of another person would inhibit the gestures that should accompany composition (10. 3. 19–21).

In addition to spatial requirements, poetic inspiration makes temporal demands, which medieval Latin poets could not help but recognize. A direct-minded soul like Aldhelm urged his students to remember that practice makes perfect. As he stressed, the craft of versification is better learned from studying metrics than from the spring or mount of the Muses.[8]

Yet where poetry is concerned, spending long stretches of time has never guaranteed good results. To quote Quintilian once again (10. 3. 15):

ad profectum enim opus est studio, non indignatione. Ut possimus autem scribere etiam plura et celerius, non exercitatio modo praestabit, in qua sine dubio multum est, sed etiam ratio.

For to make any real advance we need study, not self-accusation. And it is not merely practice that will enable us to write at greater length and with increased fluency, although doubtless practice is most important. We need judgement as well.

As Quintilian continues, he contrasts a romantic view of achieving inspiration with a highly rational and practical one:

si non resupini spectantesque tectum et cogitationem murmure agitantes expectaverimus quid obveniat, <sed> quid res poscat, quid personam deceat, quod sit tempus, qui iudicis animus intuiti humano quodam

[8] *De metris et enigmatibus ac pedum regulis* 9 (Ehwald, p. 78, ll. 10–16).

modo ad scribendum accesserimus. Sic nobis et initia et quae sequntur natura ipsa praescribit.

So long as we do not lie back with eyes turned up to the ceiling, trying to fire our imagination by muttering to ourselves, in the hope that something will present itself, but turn our thoughts to consider what the circumstances of the case demand, what suits the characters involved, what is the nature of the occasion and the temper of the judge, we shall acquire the power of writing by rational means. It is thus that nature herself bids us begin and pursue our studies once well begun.

We may wonder whether Quintilian's portrayal of a would-be orator with writer's block corresponds more closely to the modern image of thwarted writers surrounded by balls of crumpled typewriter paper, to Celtic traditions of bards lying in darkened huts, or to some other notion of composition.

Because many writers live in an imperfect world in which the daytime offers few expanses of uninterrupted time, they often resort to the night for the solitude that they require.[9] Callimachus, who designates the practice by the word *agrypnia*, seems to have been the earliest ancient writer to stress that he worked into the night. But Roman writers were likelier than their Greek counterparts to stress their nocturnal diligence. Some of them professed that night-time labours were necessary to compensate for their paucity of natural gifts,[10] while others saw a connection between the sleeplessness of lyric poets and that of the lovers they described.[11] Whatever the reasons, Latin writers developed an entire lexicon of terms to convey the special nature of night-time composition.[12] Thus their writings were produced by toiling by lamplight (*lucubrare*, 'to burn the midnight oil') or, to describe the same phenomenon slightly differently, by staying awake while normal people slumbered (*vigilare* and *vigilia*). Juvenal writes of sleepless nights in cramped quarters: 'vigilataque proelia dele, / qui facis in parva sublimia carmina cella' (*Satire* 7. 27–8). Writers called the results of these night-time studies *lucubrationes* or *lucubratiunculae*. Indeed, they even took the titles of their works from the nights that they spent in writing, such as Aulus Gellius' famous *Noctes Atticae*. Taking the concept of nocturnal diligence

[9] For the information in this paragraph I am indebted to Janson, 97–8 and 144–5.

[10] A counter-tradition that presented sleep as a precursor to composition seems slimly attested: see Horace, *Epistles* 2. 2. 78 and *Odes* 3. 4.

[11] See Thomas, 195–206.

[12] For the definitions of these words I have relied upon the *Oxford Latin Dictionary*.

to a new extreme, Martianus Capella hypostatized it as an allegorical being, Agrypnia, one of Lady Philology's servants.[13]

Of course, the habit of night-time composition was not restricted to medieval poets familiar with the Roman theories of *agrypnia* and *lucubratio*. Sometimes it was a native practice that pre-existed classical influence. For instance, in Celtic countries both storytelling and the composition of bardic poetry took place in the dark, the latter in the strictest solitude.[14] In other cases the habit of night-time composition could have resulted from a desire to demonstrate Christian *vigilance* in a very real sense. Finally, the choice of night could have been an idiosyncrasy that a poet developed without any outside suggestion whatsoever; there have doubtless always been night-owls among poets.

Because the brew of possible classical, local, Christian, and individual influences is thick, early medieval occurrences of night-time composition cannot be traced to a single source. A case in point is the famous scene in Bede's *Ecclesiastical History of the English People* (4. 24) in which Caedmon (the first English Christian poet) leaves a feast to avoid participating in singing, settles down for the night, and has a dream in which he is inspired to sing a song in praise of God the Creator. Should Caedmon's sudden access of poetic ability be viewed as a re-enactment of classical *lucubratio*, a manifestation of Christianizing *ruminatio*, an innovative application of compositional techniques current in oral tradition, a peculiarity of an individual poet, or simply as a gift of God?[15] And there is no reason to see only one of these processes at work, when all four could have been involved.

A passage toward the end of Cynewulf's *Elene* raises similar questions to those posed by Bede's portrayal of Caedmon (1236–42):

Đus ic frod ond fus þurh þæt fæcne hus
wordcræftum wæf ond wondrum læs
þragum þreodude ond geþanc reodode
nihtes nearwe. Nysse ic gearwe
be ðære rode riht ær me rumran geþeaht
þurh ða mæran miht on modes þeaht
wisdom onwreah.

[13] *De Nuptiis Philologiae et Mercurii* 2. 112 and 2. 145.

[14] On storytelling see Mac Cana, 3–5; on bardic poetry, Bergin, 6–10. It is an interesting coincidence that some of the procedures followed by bards as they composed resemble the procedures condemned by Quintilian.

[15] See Fritz (1969) and (1974). Bede himself takes the last-mentioned view: Caedmon learned the craft of poetry 'non ab hominibus, neque per hominem institutus . . . sed divinitus adiutus gratis canendi donum accepit'.

> Thus, I, aged and about to depart hence because of this frail body, have woven the art of words and have wondrously gathered my matter, have pondered at times and sifted my thought in the anguish of the night. I knew not clearly the truth about the cross till wisdom by its glorious strength revealed to the thought of my mind a larger view.

Are the nights of this poet devoted to poetry because he is old and insomniac? Or are they convenient because he is a monk, often awake to participate in prayer and song, and often anguished with thoughts of the afterlife? Or is he directing to Christian ends a compositional technique that was closely associated with Old English poetry in oral tradition? However strong the temptations, we cannot reduce the complex forces behind the night-time composition of either Caedmon or Cynewulf to a simplex motivation. We must be equally chary of being reductionist when we encounter medieval Latin poets such as Odo of Cluny and Bernard of Morlas when they seem to have engaged in meditative composition by night.[16] Because such Latin words as *meditatio* figure prominently in their descriptions, a combination of *ruminatio* and vigilance seems undeniably involved. But we are not in a position to rule out other influences. For example, because Odo was well versed in Virgil and other Roman poets, it is conceivable that his night-time meditations emulated (either consciously or not) classical *vigiliae* and *lucubrationes*.

Even when aspiring poets had found a place and time auspicious to the composition of verse, sometimes they needed additional inspiration. Although they could seek the help and prayers of the dedicatee or the audience, they often opted instead to invoke one or more divine agents for inspiration in their meditations or *musings*.[17] The poets were likely to place such invocations of divine aid at predictable junctures in poetic compositions.[18] As Quintilian observed, invocations are common not just in programmatic statements at the commencement of works, but

[16] John of Salerno describes Odo's habit of praying at night and seems to suggest that he also wrote during or after these prayer sessions: see *Vita Odonis* I. 14 (PL 133. 49 D). For the theory that Odo's *Occupatio* resulted from such meditative composition see Jan M. Ziolkowski, 'The *Occupatio* by Odo of Cluny: A Poetic Manifesto of Monasticism in the 10th Century', forthcoming.

Bernard of Morlas speaks of his hesitancy to write 'cumque in meditatione mea non paucos dies et noctes exardesceret ignis zeli, tandem accinxi me et locutus sum in lingua mea quod animo conceptum diu celaveram apud me' (Hoskier, pp. xxxvi–xxxvii).

[17] Janson, 144–5.

[18] See Wallner.

also in preparing for particularly important or difficult passages.[19] Besides securing divine support for the poet, invocations adorn the narrative and highlight significant moments in it.[20] Finally, invocations win the goodwill of the audience by emphasizing their devotion to God or the gods and by exciting their interest in the importance of their themes.[21]

The gods most commonly invoked for inspiration and knowledge in classical Greek and Latin poetry were the Muses.[22] It is upon them that Homer calls in the opening lines of *Iliad* and *Odyssey* for inspiration and later in the *Iliad* for knowledge (2. 484–93). Virgil entreats them for inspiration in the *Aeneid* (1. 8), for the knowledge of cosmic laws in the *Georgics* (2. 475–82). As the last example suggests, the Muses were associated not just with epic, but also with pastoral and didactic.[23]

Any convention so well-established as the invocation of the Muses almost inevitably provoked parodic and ironic treatment.[24] The invocation of the Muses is parodied as early as Propertius, who professes (2. 1. 3–4):[25]

> non haec Calliope, non haec mihi cantat Apollo.
> ingenium nobis ipsa puella facit.
>
> Not Calliope, not Apollo sings these words to me;
> my girlfriend herself brings me inspiration.

Horace parodies the customary epic invocation of the Muses in the *Satires* (1. 5. 51–4), while Ovid treats the Muses with characteristic irony in the *Ars amatoria* (1. 27–30 and 2. 704) and *Remedia amoris* (361–2 and 387–8).

The self-conscious irony of the Hellenistic poets and their successors toward the Muses was possible among Christian poets only long after the Muses had ceased to be regarded as threatening vestiges of a non-Christian culture.[26] Gradually the Muses came to be made palatable, appearing less as inspiring goddesses than as the embodiments of the individual poet's own talents or 'genius'.[27] Thereafter humour and irony in attitudes toward the

[19] 4. 4 ('Prohoemium'); cited by Curtius, *European Literature*, 232.
[20] Curtius, loc. cit.
[21] 10. 1. 48.
[22] See Murray, 89–92.
[23] Curtius, *European Literature*, 232.
[24] All but one of the examples in this paragraph are cited by Curtius, loc. cit.
[25] Cf. Tibullus 2. 1. 35.
[26] For studies of the Muses in Christian poetry see Curtius, *European Literature* and 'Die Musen im Mittelalter'; Heitmann; and Klopsch, 26–33.
[27] Klopsch cites Walahfrid Strabo (*MGH Poetae*, ii. 370) and Hucbald's *De calvis* (*MGH Poetae*, iv. 267 ff.). Similar invocations of the Muses are found in Walter of Châtillon, *Alexandreis* 1. 1–5, and Dante, *Purgatorio* 1. 7–12.

Muses reappear. For example, one of the *Carmina Burana* contains not an invocation of the nine Muses, but an oath by them, Jove, Apollo, Mars, and Cupid that the poet has never been nor ever will be unfaithful to his lover.[28]

But many Christian thinkers, particularly in the early centuries, rejected the notion of invoking the Muses and Apollo for inspiration. The poems of Paulinus of Nola contain several such renunciations, of which the following two are well-known and typical:

Negant Camenis nec patent Apollini
Dicata Christo pectora.[29]

Hearts dedicated to Christ refuse the Muses and are shut to Apollo.

Non ego Castalidas, vatum phantasmata, Musas
nec surdum Aonia Phoebum de rupe ciebo;
carminis incentor Christus mihi, munere Christi
audeo peccator sanctum et caelestia fari.[30]

I do not call forth the Muses of Castalia, those phantoms of poets, nor deaf Apollo of Mount Helicon; it is Christ who elicits my song, and though a sinner I dare to speak of the holy and heavenly through the gift of Christ.

Such rejection of the Muses and Apollo remained in fashion long into the Middle Ages.[31] Yet although they sometimes discarded the practice of invoking the Muses and Apollo, Christian Latin poets remained devoted to the principle of invocation as a means of achieving inspiration; only now they turned to suitable Christian substitutes in their quest for divine assistance: to Christ, to saints, but above all to the Holy Spirit.[32]

The most famous early replacement of the Muses and Apollo appears in the preface to the 'Bible epic' of Juvencus. The poet exclaims (*Praefatio* 25–7):[33]

Ergo age! sanctificus adsit mihi carminis auctor,
spiritus, et puro mentem riget amne canentis
dulcis Iordanis, ut Christo digna loquamur.

Come on then! Let the creator of song, the sanctifying spirit, be at my side and let it water my heart from the pure flow of the sweet Jordan that I may speak words worthy of Christ.

[28] Poem 117 (*incipit* 'Lingua mendax et dolosa', ed. Hilka-Schumann, i. 2, 192–3).
[29] *Carmina* 10. 21–22; cited by Curtius, *European Literature*, 235.
[30] *Carmina* 15. 30–33; cited by Klopsch, 26.
[31] Klopsch, 26–7.
[32] See Klopsch, 20–30; Roberts, 124 n. 7.
[33] Cited by Curtius, 235, Klopsch, 22.

Rather than calling upon the Muses and Apollo, Juvencus has recourse to the Holy Spirit. Rather than drinking from the Muses' spring, he asks to be sprinkled with water from the river Jordan.

From this point onward invocations of the whole godhead or the Holy Spirit became a standard component of exordia in Christian Latin poems.[34] Eventually these invocations became so stylized that they constitute self-contained hymns.[35] These invocations of the Holy Spirit left their mark on vernacular as well as medieval Latin poetry. To cite two examples from early German Bible poetry, there are the *Lob Salomons* (1. 1–10)[36]

Inclita lux mundi,
du dir habis in dinir kundi
erdin undi lufti
unde alli himilcrefti,
du sendi mir zi mundi,
daz ich eddilichin deil muzzi kundin
di gebi vili sconi,
di du deti Salomoni,
di manicfaltin wisheit:
ubir dich mendit du cristinheit.

Inclita lux mundi,
who in your discernment hold
the earth and air
and all the powers of heaven:
bestow upon my mouth
that I may make known some part
of the many beautiful gifts
that you gave to Solomon,
the manifold wisdom:
through you Christianity exults.

and the *Exodus*, where an invocation of God and appeal for inspiration are at their most plaintive (17–20 and 29–34):[37]

dû gib mir dînen wîstûm
daz ich muge wandilôn
mit tûtiskeme munde
der latînsken zungen. . . .

[34] In addition to the passages cited by Curtius, see Sidonius Apollinaris, *Carmina* 16. 5–70 and *Eupolemius* 5–12.

[35] As examples, Klopsch, 22, refers to Marius Victor, *Alethias* and Sedulius, *Carmen paschale* 1. 60–102. The same process of stylization probably lies beyond some of the Homeric hymns.

[36] Cited by Haug, 48. My translation.

[37] Cited by Haug, 49. My translation.

nû sende mir sanctum
spiritum paraclitum,
der mîn gebende lôse,
sô wil ich gerne chôsen;
der heilige geist dîn
ordene die rede mîn.

Give me your wisdom,
so that I can translate
into German
the Latin tongue. . . .
Now send me the *spiritus*
sanctus, the paraclete,
to loose my bonds,
so I will speak readily.
May the Holy Spirit—yours—
direct my speech.

The origins of such invocations are distinctly visible in both of these passages. Both proclaim their indebtedness to the Latin culture of the Church, the one in the opening vocative and later dative *Salomoni*, the other in the outright plea for aid in translating Latin that is directed to the *sanctus spiritus paraclitus*—the Holy Spirit. Making overt what was covert in the Old English *Elene*, both call upon the Holy Spirit specifically as the wisdom of the godhead, the agent through which God the Creator can impart verbal powers to the mouth of a poet. The Latin tag with which the first passage commences also draws attention to the illuminatory capacities of the Holy Spirit.

Although sometimes poets invoked God the Creator as they sought the power to interpret and reveal the eternal truth behind temporal affairs, most often they requested inspiration from the Holy Spirit.[38] Medieval poets had at least three compelling reasons for connecting the Holy Spirit with inspiration: first, because the *Spiritus Sanctus* was easily coordinated with the widespread conception of in*spir*ation as a breathing that comes upon a poet; second, because the Holy Spirit was associated with an intensely verbal form of spiritual inspiration; and third, because the Holy Spirit had been shown repeatedly in the Bible to be capable of filling men with the powers of poetry and prophecy. Each of these three associations of the Holy Spirit merits closer examination.

[38] See Haug, 47 n. 3.

The conception of inspiration as a breathing that comes upon a poet or as a drawing in of spiritual breath is embedded in many Indo-European languages. At two extremes of the geographical spectrum in the most archaic evidence, both Old Irish and Vedic contain words in which poetic art is linked metaphorically with blowing, breath, and wind.[39]

In ancient Greek literature the belief in breath as a medium of contact between gods and men is common;[40] but there are few metaphors that connect such breathing with poetic inspiration in particular. In an apparently isolated passage, Hesiod uses the verb *enepneusan* ('in-breathed, in-*spired*') to describe how the Muses inspired him with song as he was shepherding his lambs under Helicon.[41] Only after the arrival of Christianity do breathing metaphors spread in Greek. According to Clement of Alexandria, who purports to be quoting the words of Democritus (born *c.*460 BC): 'The things a poet writes . . . with *enthousiasmos* and holy spirit are all very beautiful' (*met' enthousiasmoû kaì hieroû pneúmatos*).[42] Taken loosely, this supposedly Democritean conception of divine possession (*enthousiasmos*) seems to anticipate the New Testament notion of divine in-spiration, expressed through the word *theopneustos* (2 Timothy 3: 16). The resemblance of the two conceptions is brought home through the wording *hieròn pneûma*, which is but a hair's breadth from *pneûma hágion*, a phrase by which the Holy Spirit is designated in the Greek New Testament (e.g. John 20: 22).[43]

Christian Latin writers appropriated the traditional metaphor of blowing and breath because it suited the terminology of the Holy Spirit. In so doing they took over the vocabulary that had been used repeatedly in classical Latin to describe the means by which people became inspired, in either spirit or writing. Thus they adopted words such as *afflatus* (*TLL* 1. 1229. 22–48), *afflo* (*TLL* 1. 1241. 7–32), *concitatio*, *concito*, and *instinctus*. But Christian Latin writers did not accept the entire pre-Christian vocabulary without modification. Through inventing new words and giving special meanings to old ones Christian Latin writers managed to mark off their new notion of inspiration from the pagan one of earlier writers. Whereas *aspiratio* had been used solely to describe

[39] Watkins, 215–17. [40] Onians, 50–6; Smith, 418–21.

[41] *Theogony* 31. Cf. John 20: 22 *enephúsesan*. [42] Fr. B 18.

[43] The authenticity of the phrase *kaì hieroû pneúmatos* has been questioned: see Guthrie, ii. 477 n. 2. The phrase has been explained plausibly as a gloss on *met' enthousiasmoû*: see Fronmüller, 21–2.

the movement of physical breath or winds and to designate letters and signs for rough breathing in grammar, among Christian Latin writers it indicated divine inspiration.[44] Whereas *flamen* had been used to describe winds and similar blowings, in Christian Latin it could also mean either inspiration through the Holy Spirit or the Holy Spirit itself.[45]

The most important of their lexical innovations was the word *inspiratio* itself. Pre-Christian Latin poets had used not *inspiro*, but *adspiro* to describe the breathing-like action by which divine beings filled poets with spirit. For instance, the verb *aspiro* appears prominently in invocations of divine aid in both Virgil and Ovid:

vos, o Calliope, precor, adspirate canenti . . . (*Aen.* 9. 525)
I call upon you, O Calliope: inspire me as I sing . . .

Di coeptis (nam vos mutastis et illas)
adspirate meis, primaque ab origine mundi
ad mea perpetuum deducite tempora carmen. (*Metam.* 1. 2–4)

O Gods, inspire me in my undertakings, for you caused these changes, and guide my song steadily from the very beginning of the world to my own times.

Although derivatives of *inspiro* are well known in many modern European languages, the figurative and spiritual senses of this verb and its derivatives *inspiraculum*, *inspiramen*, *inspiramentum*, *inspirantia*, and *inspiratio* came into being only in the third and later centuries of the Christian era.[46] Because the Christians viewed the Holy Spirit as the breath that is breathed into the faithful, they considered the Holy Spirit the *inspirator par excellence* (*TLL* 7/1. 1959. 61–72).

At different moments in the Old Testament the *spiritus Dei* is presented as a creative, enlivening power (Genesis 2: 1, Job 33: 4). Already equated with the Word of God in the Psalms (32: 6), the *Spiritus Sanctus* gradually became connected not only with spiritual and artistic inspiration in general, but also with a specifically verbal form of inspiration. The foremost poet-prophet was David, whose obligation to the Holy Spirit was a matter of record in both the Testaments (2 Samuel 23: 2, Mark 12: 36, and Acts 1: 16); but David was by no means alone in receiving a verbal gift from this source. In Zechariah the people refuse to 'hear the law, and the words which the Lord of hosts sent in his spirit by the hand of the

[44] *TLL* 2. 838. 21–839. 72.
[45] *TLL* 6/1. 861. 19–22; Blaise, 152 and 359.
[46] *TLL* 7/1. 1959. 24–59; Souter, 211.

former prophets' (7: 12). Indeed, the Holy Spirit proved its ability to fill with the powers of poetry and prophecy not only the prophets of Israel but even Balaam and his ass—a man and an animal outside the people of Israel (Numbers 22–4).

The tendency of the Old Testament to present the Holy Spirit as a verbal force was strengthened in the New Testament. 'In the beginning was the Word, and the Word was with God, and the Word was God': the very first verse in John asserts that major events of salvation as seen from Christian eyes—namely, creation and the birth of Christ—were verbal acts; and these verbal acts were understood to have involved the operation of the Holy Spirit. The New Testament is God-speech in several senses, in that it records not only the acts of Christ, but also the words of Christ, the acts of God's verbal force, and even the words of the Holy Spirit (Hebrews 3: 7). Accordingly, it was appropriate for medieval artists to represent Gospel-writers and the composers of other sacred texts with angelic or dove-like figures seated on their shoulders or hovering in the air above them, dictating to them and inspiring them.

Because the Holy Spirit was capable of verbal acts, medieval poets had ample reason to invoke it for inspiration. At the beginning of the *De vulgari eloquentia* Dante provided the most famous but not the earliest example of Christ invoked as the Word: 'Verbo aspirante de coelis'. To take one typical example, the author of the *Vita beati Leudegarii martyris* also invoked Christ with the same focus (1. 34–44):[47]

> Linguae clustella resolve,
> Linguam qui dederas dederasque exordia vitae.
> Barbara sed reboet quamvis compressa labellis;
> Namque potes, poteras quondam qui guttura aselli
> Humanis formans labiis similare loquendo.
> Mens mea iam gestit tanti depromere laudes
> Vatis, sed graciles vires torpore tepescunt.
> Non adeo sum desperans, quin omnia possis;
> Gnaro nam memini scriptum pandente propheta:
> 'Ora dehisce libens, ego donis illa replebo.'

Undo the locks of the tongue [the Word], you who granted the tongue [the Word] and granted the beginnings of life. Let the uncouth tongue, although pressed by the lips, resound; for surely you have this power,

[47] Traube's notes list many parallels, most of which are also noted in Curtius, *European Literature*, 237 and 239. The passage is modelled closely upon Aldhelm, *Carmen de virginitate* 56–80 (tr. Lapidge and Rosier, pp. 104–5).

you who once had the power to render the throat of an ass like human lips in speech. My mind now wishes to produce praise of a great prophet, but my meagre potential lies slack and feckless. Yet I do not despair, since you are capable of anything; for I recall what a wise prophet once wrote: 'Open thy mouth wide and willingly, and I will fill it with gifts.'

This passage contains two thoughts that had been commonplace in invocations of divine aid by Christian Latin writers since late antiquity. The first is the hope of being possessed by the Holy Spirit like Balaam or his ass;[48] the other is the application of Psalm 80: 11 'Open thy mouth wide, and I will fill it' ('dilata os tuum et implebo illud') to convey an image of the Christian writer as an orifice that God fills by *in-spiring*.[49] Because this image renders human beings into mouthpieces of divine will, it is not surprising to encounter a ninth-century legend that Gregorian chant originated when the Holy Spirit whispered the melodies into Gregory the Great's ears.[50]

Doctrines of inspiration often presuppose that a person is empty until being filled with inspiration. Because the filling can result from the intervention of an external agent, the inspired person sometimes bears a strong resemblance to a man possessed. Such ecstatic possession, similar to experiences that shamans and magicians are reputed to undergo, seems to take place in the Old Testament when Saul 'stripped himself also of his garments, and prophesied with the rest before Samuel, and lay down naked all that day and night' (1 Samuel 19: 20–4). In classical philosophy and literature the inspiration of the poets was often equated explicitly with the frenzy of the possessed, most notably in Plato's dialogues.[51]

Although the relevant passages of Plato were not available in the West during the Middle Ages, the equation of poets with frenzied devotees of gods was known through its appearance in classical Latin literary theory. Horace ridicules this equation, whereas Cicero seems to take it seriously in asserting that the only good poet is the one inspired with frenzy.[52] Pliny states as a principle that 'Poetis furere concessum est' (*Epistles* 7. 4. 10 'poets

[48] For references to Orientius and Braulio see Curtius, *European Literature*, 236–7; Klopsch, 23–4.

[49] Janson, 144.

[50] Fritz (1969), 335.

[51] *Ion* 533 E; *Laws* 719 C; *Phaedrus* 245 A. For other references see Nisbet and Hubbard on Horace, *Odes* 2. 19; Russell, 74; and, above all, Tigerstedt (1969) and (1970).

[52] Horace, *Ars poetica* 295–8 and 455; Cicero, *De oratore* 2. 46. 194 (cf. *De divinatione* 1. 36. 80). See Russell, 72.

are allowed to talk wildly'). In addition to these more or less systematic treatments, the concept of *furor poeticus* is touched upon casually in poetry: Horace refers casually to 'amabilis insania' in *Odes* 3. 4. 5, while Ovid speaks of being inspired by the deity (*Fasti* 6. 5; *Ex Ponto* 3. 4. 93 and 4. 2. 25). The concerns of the literary theorists and poets did not fall on deaf ears, to judge by the continued mentions of these topics in Silver Latin and Late Latin poetry.[53]

The concept of *furor poeticus* would have been known through channels outside the mainstream of classical literature. Through a false etymology the encyclopedist Isidore of Seville (*c*.570–636) connects the poet (*vates*) with the prophet and madman:

Vates a vi mentis appellatos Varro auctor est; vel a viendis carminibus, id est flectendis, hoc est modulandis: et proinde poetae Latine vates olim, scripta eorum vaticinia dicebantur, quod vi quadam et quasi vesania in scribendo commoverentur; vel quod modis verba conecterent, viere antiquis pro vincire ponentibus. Etiam per furorem divini eodem erant nomine . . .[54]

Varro held that *vates* are so called from 'force of mind' (*vis mentis*), or from weaving songs, that is plaiting them, which is to say setting them to music; and accordingly, poets were formerly called *vates* in Latin and their writings *vaticinia*, because in writing they were moved by a certain force (*vis*) and as if by madness; or because they join words in measures, since people in earlier times employed *vieo* in place of *vincio*. They also had the name *vates* on account of their divinely inspired rage . . .

Vates a vi mentis appellatos, cuius significatio multiplex est. Nam modo sacerdotem, modo prophetam significat, modo poetam.[55]

The *vates* are so called from 'force of mind' (*vis mentis*), the significance of which is multiple. For the word now means priest, now prophet, and now poet.

Remaining beyond the pale of linguistic truth Isidore etymologizes the word *carmen* as follows: 'Carmen vocatur . . . quod qui illa canerent carere mentem existimabantur' ('A song is so called because those who sang them were believed to have lost their minds').[56] Isidore's etymology has a clear resonance in Modoin's 'Nonnulli adfirmant etiam insanire poetas, / Carmina dum statuunt mente carere sua' ('Some people assert that poets are out of their minds, / That they lose their minds while creating poems').[57]

[53] For references see Curtius, *European Literature*, 474.

[54] *Etymologiae* 8. 7. 3. On the nuances of the word *vates* in classical Latin, see Newman.

[55] Ibid. 7. 12. 15.

[56] Ibid. 1. 39. 4.

[57] *MGH Poetae*, i. 570, ll. 23–4. Cited by Curtius, *European Literature*, 474.

A more accurate etymology would connect *vates* not with madness but with insight or second sight. The word belongs to a complex that includes such Germanic names and words as: óðinn/Woden/Wotan, Germanic god of poetry, writing, magic, and wisdom; Old English *wōðbora* 'seer, prophet, singer'; and Old Norse *óðr* 'mind, wit, soul; song, poetry' (to be compared with the adjective meaning 'mad, frantic').[58] But however risible Isidore's derivation of *vates* from *vi mentis* may be to a linguist, the sixth-century Spanish scholar and his source perceived one of the fundamental truths about the original nature of the *vates*: that he was understood to be possessed.

The spirits that induced inspiration in poets were not uniformly divine. Because the mouths of poets were not always filled with the Holy Spirit alone as they composed verse, a treatment of inspiration in ancient and medieval times would not be complete without consideration of alcohol. As Ernst Robert Curtius advised as a remedy against the staleness of the Muses in modern poetry: 'Better—as Samuel Butler did in his *Hudibras* (1663)—to invoke a jug of beer, which perhaps has inspired more poetry and prose than all the waters of Hippocrene and Helicon.'[59]

When did consumption of alcohol come to be regarded as inspirational? Already in Archilochus wine is employed to aid composition. Whether or not the practice of drinking to facilitate composition was common among the early poets cannot be known, but a later tradition held that the best poets were at their best when they had wine to drink.[60] This creed attained the status of a dictum in Greek lines attributed to Cratinus.[61] In Roman literature the belief became especially well rooted in Horace:

Prisco si credis, Maecenas docte, Cratino,
nulla placere diu nec vivere carmina possunt,
quae scribuntur aquae potoribus.[62]

If you trust in old Cratînus, my learned Maecenas,
no poems can please long or live,
which are written by water-drinkers.

Ennius ipse pater numquam nisi potus ad arma
prosiluit dicenda.[63]

Even Father Ennius never leapt forth to tell of arms unless drunk.

[58] Different etymologies have been proposed for different senses of *óðr*: see de Vries, 416. On the whole complex see Chadwick, 10 and 13.

[59] Curtius, *European Literature*, 245.

[60] Knox, 109.

[61] Fr. 199 K.

[62] *Epistles* 1. 19. 1–3.

[63] Ibid. 1. 19. 7–8.

fecundi calices quem non fecere disertum?[64]
Whom have full drinkbowls not made eloquent?

Closely related to the belief that wine is inspirational was the belief that Bacchus is pre-eminently a poet's god. This belief in Bacchus took shape because in one of his numerous guises he was a god of vegetation who suffered death but came to life again as a god of wine, able to loosen care (Lyaeus) and inspire creativity in music and poetry. (As such Bacchus anticipated a few major features of John Barleycorn, the English and Scottish personification of the grain from which alcoholic drinks are made; for John Barleycorn is a victim who dies so that drinkers may enjoy his blood.)[65] Bacchus was mentioned as a patron deity of poets repeatedly by Augustan poets such as Propertius and Horace.[66]

Poetic inspiration and alcohol are sometimes coupled in Carolingian poetry, as in a verse dialogue by Mico of St Riquier in which Mico converses with his Muse. Here the classical idea that the consumption of alcohol promotes inspiration is turned around, inasmuch as Mico's Muse requests beer and wine as rewards *after* she has sung at Christian festivals: inspiration promotes the consumption of alcohol.[67] But later in the Middle Ages the customary classical progression returned with a vengeance, as a survey of Latin proverbial sayings will easily verify.[68]

The most famous passage on drunken inspiration in medieval Latin literature comes from the twelfth century, in the Archpoet's magnificent 'Confession' (10. 13–19):

13 Poculis accenditur animi lucerna;
cor inbutum nectare volat ad superna.
mihi sapit dulcius vinum de taberna
quam quod aqua miscuit presulis pincerna.

64 *Epistles* 1. 5. 19.

65 Leach and Fried, 554.

66 Knox, 111 n. 14, cites Propertius 2. 30. 37–40, 3. 2. 9, 4. 1. 62, 4. 6. 75; see also Nisbet and Hubbard's introductory note to Horace *Odes* 2. 19. Horace describes himself as carried away by Bacchus into groves and grottoes in *Odes* 3. 25, and deals ironically with the inspiration of Bacchus in *Epistles* 1. 19. 3–4 and 2. 2. 78.

67 *Carmina centulensia* 16. 75–6 and 17. 33–4.

68 A few examples from Walther: 'Cum bene sum potus, tunc versibus effluo totus; / Cum sicco, siccor, nec in hic nec in hec (hac?) nec in hoc cor' (no. 4034); 'Cum bibo vinum, versum facio leoninum' (no. 4039: cf. 6660); 'Dum bene sum potus, tunc verbis affluo totus; / Si sum ieiunus, tunc sum de pauperibus unus' (no. 6416); 'Dum bibo vinum, loquitur mea lingua Latinum; / Dum bibo cervisiam, tunc perdo philosophiam' (no. 6427); 'Qui bona vina bibunt, vates bona carmina scribunt' (no. 23883).

14 Loca vitant publica quidam poetarum
et secretas eligunt sedes latebrarum,
student instant vigilant nec laborant parum
et vix tandem reddere possunt opus clarum.

15 Ieiunant et abstinent poetarum chori,
vitant rixas publicas et tumultus fori
et, ut opus faciant quod non possint mori,
moriuntur studio subditi labori.

16 Unicuique proprium dat Natura munus:
ego nunquam potui scribere ieiunus;
me ieiunum vincere posset puer unus.
sitim et ieiunium odi tamquam funus.

17 Unicuique proprium dat Natura donum:
ego versus faciens bibo vinum bonum
et quod habent purius dolia cauponum;
tale vinum generat copiam sermonum.

18 Tales versus facio, quale vinum bibo;
nihil possum facere nisi sumpto cibo;
nihil valent penitus que ieiunus scribo;
Nasonem post calicem carmine preibo.

19 Mihi nunquam spiritus poetrie datur,
nisi prius fuerit venter bene satur;
dum in arce cerebri Bachus dominatur,
in me Phebus irruit et miranda fatur.

13 The lamp of the spirit is lit by cups of wine;
a heart steeped in nectar soars to heaven.
Wine from a tavern tastes sweeter to me than
what the bishop's wine-steward has watered down.

14 Certain poets avoid public places
and opt for secluded hiding-places;
they study, persevere, keep vigil, and toil not a little,
and can still scarcely produce a great poem.

15 The bands of poets go hungry and do without,
avoid public hubbub and the commotion of the market-place,
and, to produce a work so that they will live forever,
they die of ambition, overworked.

16 Nature gives each person a special gift:
I never could write verse while fasting;
a mere boy could outdo me when I fast.
Thirst and hunger are death to me: I hate them.

17 Nature gives each person a special present:
I make verse while drinking good wine—
the purest that the vats of the innkeepers contain.
Wine of this sort makes me eloquent.

18 The quality of my verse-making depends on my wine-drinking;
and I can accomplish nothing unless I have eaten—
what I write when hungry is worth nothing.
But after a glass I will outdo Ovid in song.

19 The spirit of poetry is never granted me,
unless my belly has first been well filled; while Bacchus
holds sway in the control-centre of my brain, Apollo
rushes upon me and makes marvellous pronouncements.

These seven stanzas bring together with seeming effortlessness many strands of Roman doctrine about inspiration. The Archpoet alludes to poetic illumination and poetic flight before contrasting wine and water as agents of inspiration. Then he appraises the value of seclusion, implying that the hallowed techniques of withdrawal from the world and *lucubratio* do not guarantee poetic success. Thereafter he focuses his eye on wine and food, which he claims to need before he can compose; because for him wine/Bacchus is the indispensable precursor to poetic wisdom/Apollo.

The Archpoet's hearty appropriation of ancient approaches to inspiration in his 'Confession' is particularly amusing in the light of an earlier poem, in which he professed tongue-in-cheek to reject Apollo and the Muses for reasons of religion (9. 7):

Filius ecclesie fidem sequor sanam,
contempno gentilium falsitatem vanam;
unde iam non invoco Febum vel Dianam
nec a Musis postulo linguam Tullianam.

A good son of the Church, I adhere to sound faith,
and I spurn the vain falseness of pagans;
for this reason I do not invoke Apollo or Diana,
and I do not seek the eloquence of Cicero from the Muses.

If the mechanisms that poets construct to account for their creative processes can be called playful, then the play reaches a high pitch of sophistication and self-consciousness in such poems as this, where classical conventions of inspiration are tested, adapted, and revitalized through interaction with the Christianity of the medieval Latin poets.

BIBLIOGRAPHY

PRIMARY

Aldhelm, *Opera*, ed. Rudolph Ehwald, *MGH Auctores Antiquissimi*, xv (Berlin, 1919).
The poetic Works, tr. M. Lapidge and J. L. Rosier (Cambridge, 1985).
Archpoet, *Die Gedichte des Archipoeta*, ed. Heinrich Watenphul and Heinrich Krefeld (Heidelberg, 1958).
Bede, *Ecclesiastical History of the English People*, ed. and tr. Bertram Colgrave and R. A. B. Mynors (Oxford Medieval Texts: Oxford, 1969).
A History of the English Church and People, tr. Leo Sherley-Price, rev. R. E. Latham (Baltimore, 1968).
Bernard of Morlas (Bernard of Cluny), *De contemptu mundi*, ed. H. C. Hoskier (London, 1929).
Bible, Latin. *Biblia sacra iuxta vulgatam versionem*, ed. Robert Weber (2 vols., 2nd edn., Stuttgart, 1975).
The Holy Bible: Douay Rheims Version (Baltimore, 1899).
Carmina Burana, ed. Alfons Hilka and Otto Schumann, I. 2 (2nd imp., Heidelberg, 1971).
Cynewulf, *Elene*, ed. George Philip Krapp, *The Vercelli Book* (Anglo-Saxon Poetic Records, 2: New York, 1932), 66–102.
tr. R. K. Gordon, *Anglo-Saxon Poetry* (London, 1926).
Democritus, ed. and tr. G. Lanata, *Poetica pre-Platonica* (Florence, 1963).
Eupolemius, *Das Bibelgedicht*, ed. Karl Manitius, *MGH Quellen zur Geistesgeschichte der Mittelalters*, ix (Weimar, 1973).
Exodus, ed. Edgar Papp, *Das altdeutsche Exodus* (Medium Aevum Philologische Studien, 16: Munich, 1968).
Horace, *Opera*, ed. Edward C. Wickham, rev. H. W. Garrod (Oxford Classical Texts: Oxford, 1911).
Isidore, *Etymologiarum sive originum libri XX*, ed. W. M. Lindsay (2 vols., Oxford Classical Texts: Oxford, 1911).
John of Salerno, *Vita Odonis*, ed. Migne, PL 133. 43–86.
tr. G. Sitwell, *St. Odo of Cluny. Being the Life of St. Odo of Cluny by John of Salerno and the Life of St. Gerald of Aurillac by St. Odo* (London-New York, 1958), 1–87.
Lob Salomons, ed. Albert Waag and Werner Schröder, *Kleinere deutsche Gedichte des 11. und 12. Jahrhunderts* (Tübingen, 1972), 46–55.
Martianus Capella, *De nuptiis Philologiae et Mercurii*, ed. Adolfus Dick, with additions by Jean Préaux (Bibliotheca Teubneriana, 2nd edn.: Stuttgart, 1978).
Mico of St Riquier, *Carmina centulensia*, ed. Ludwig Traube, *MGH Poetae Latini Aevi Carolini*, iii (Berlin, 1886–96), 279–368.

Odo of Cluny, *Occupatio*, ed. A. Swoboda (Bibliotheca Teubneriana: Leipzig, 1900).

Ovid, *Amores*, *Medicamina faciei femineae*, *Ars amatoria*, *Remedia amoris*, ed. E. J. Kenney (Oxford Classical Texts: Oxford, 1965).

Persius, *The Satires*, tr. Guy Lee, introd. and comm. William Barr (Latin and Greek Texts, 4: Liverpool, 1987).

Plato, *Ion* and *Phaedrus*, in *The Collected Dialogues of Plato*, ed. Edith Hamilton and Huntington Cairns (Bollingen Series, 71: Princeton, 1963).

Pliny, *Letters and Panegyricus*, ed. and tr. Betty Radice (2 vols., Loeb Classical Library: Cambridge, Mass., 1969).

Propertius, *Elegies*, ed. E. A. Barber (Oxford Classical Texts: 2nd edn., Oxford, 1960).

Quintilian, *Institutio oratoria*, ed. M. Winterbottom (2 vols., Oxford Classical Texts: Oxford, 1970).

ed. and tr. H. E. Butler (4 vols., Loeb Classical Library: Cambridge, Mass., 1920–2).

Sidonius Apollinaris, *Poems and Letters*, ed. and tr. W. B. Anderson (Loeb Classical Library: Cambridge, Mass., 1936).

Tacitus, *Dialogus de oratoribus*, tr. W. Peterson, rev. M. Winterbottom, *Tacitus*, i (Loeb Classical Library: Cambridge, Mass., 1970), 217–347.

Vita Beati Leudegarii martyris, ed. Ludwig Traube, (*MGH Poetae Latini Aevi Carolini*, iii (Berlin, 1886–96), 5–37.

SECONDARY

Benham, Allen R., 'Inspiration', in *The Princeton Encyclopedia of Poetry and Poetics*, ed. Alex Preminger, Frank J. Warnke, and O. B. Hardison Jr. (Princeton, 1965), 396–8.

Bergin, Osborn, *Irish Bardic Poetry*. Foreword by D. A. Binchy, ed. by David Greene and Fergus Kelly (Dublin, 1984).

Blaise, Albert, *Le Vocabulaire latin des principaux thèmes liturgiques* (Turnhout, 1966).

Chadwick, N. Kershaw, *Poetry and Prophecy* (Cambridge, 1942).

Curtius, Ernst Robert, *European Literature and the Latin Middle Ages*, tr. William Trask (New York, 1953), 228–46 on 'The Muses' and 474–5 on 'The Poet's Divine Frenzy'.

'Die Musen im Mittelalter', *Zeitschrift für romanische Philologie*, 59 (1939), 129–88, and 63 (1943), 256–68.

Dodds, E. R., *The Greeks and the Irrational* (Berkeley, 1951; Boston, 1957).

Fritz, Donald W., 'Caedmon: A Monastic Exegete', *American Benedictine Review*, 35 (1974), 351–67.

'Caedmon: A Traditional Christian Poet', *Mediaeval Studies*, 31 (1969), 334–7.

Fronmüller, Wilhelm, *Demokrit, seine Homer-studien und Ansichten* (Inaugural Dissertation: Erlangen, 1901).

GUTHRIE, W. K. C., *A History of Greek Philosophy* (Cambridge, 1965).

HAUG, WALTER, *Literaturtheorie im deutschen Mittelalter: Von den Anfängen bis zum Ende des 13. Jahrhunderts* (Darmstadt, 1985).

HEITMANN, KLAUS, 'Boethius' Verdammung der Musen im Mittelalter', in *Renatae Litterae: Studien zum Nachleben der Antike und zur europäischen Renaissance August Buck zum 60. Geburtstag . . . dargebracht . . .*, ed. Klaus Heitmann and Eckhart Schroeder (Frankfurt, 1973), 23–49.

JANSON, TORE, *Latin Prose Prefaces: Studies in Literary Conventions* (Studia Latina Stockholmiensia, 13: Stockholm, 1964).

KAMBYLIS, ATHANASIOS, *Die Dichterweihe und ihre Symbolik: Untersuchungen zu Hesiodos, Kallimachos, Properz und Ennius* (Heidelberg, 1965).

KLOPSCH, PAUL, *Einführung in die Dichtungslehren des lateinischen Mittelalters* (Darmstadt, 1980).

KNOX, PETER E., 'Wine, Water, and Callimachean Polemics', *Harvard Studies in Classical Philology*, 89 (1985), 107–19.

LEACH, MARIA, and FRIED, JEROME, *Funk and Wagnalls Standard Dictionary of Folklore, Mythology, and Legend* (San Francisco, 1984).

MAC CANA, PROINSIAS, *The Learned Tales of Medieval Ireland* (Dublin, 1980).

MURGIA, CHARLES E., 'Pliny's Letters and the *Dialogus*', *Harvard Studies in Classical Philology*, 89 (1985), 171–206.

MURRAY, PENELOPE, 'Poetic Inspiration in Early Greece', *Journal of Hellenic Studies*, 101 (1981), 87–100.

NEWMAN, J. K., *The Concept of Vates in Augustan Poetry* (Collection Latomus, 89: Brussels, 1967).

NISBET, R. G. M., and HUBBARD, MARGARET, *A Commentary on Horace: Odes Book II* (Oxford, 1978).

ONIANS, R. B., *The Origins of European Thought about the Body, the Mind, the Soul, the World, Time, and Fate: New Interpretations of Greek, Roman and kindred evidence, also of some Basic Jewish and Christian Beliefs* (Cambridge, 1951).

Oxford Latin Dictionary, ed. P. G. W. Glare (Oxford, 1968–82).

ROBERTS, MICHAEL, *The Jeweled Style: Poetry and Poetics in Late Antiquity* (Ithaca, 1989).

RUSSELL, D. A., *Criticism in Antiquity* (Berkeley, 1981), esp. 69–83, 'The Poet and his Inspiration'.

SMITH, WESLEY D., 'So-called Possession in Pre-Christian Greece', *Transactions and Proceedings of the American Philological Association*, 96 (1965), 403–26.

SOUTER, ALEXANDER, *A Glossary of Later Latin to A.D. 600* (Oxford, 1949).

Thesaurus linguae latinae (Leipzig, 1900–). [*TLL*]

THOMAS, RICHARD F., 'New Comedy, Callimachus, and Roman Poetry', *Harvard Studies in Classical Philology*, 83 (1979), 179–206.

TIGERSTEDT, E. N., 'Plato's Idea of Poetical Inspiration', *Commentationes Humanarum Litterarum: Societas Scientiarum Fennica*, 44, no. 2 (1969), 5–76.

'"Furor Poeticus": Poetic Inspiration in Greek Literature before Democritus and Plato', *Journal of the History of Ideas*, 31 (1970), 163–78.

VRIES, JAN DE, *Altnordisches etymologisches Wörterbuch* (2nd edn., Leiden, 1977).

WALLNER, GUSTAVUS, 'De divini auxilii invocatione in carminibus epicis', *Latinitas*, 29 (1981), 263–9.

WALTHER, HANS, *Proverbia sententiaeque latinitatis medii aevi: Lateinische Sprichwörter und Sentenzen des Mittelalters in alphabetischer Anordnung* (Carmina medii aevi posterioris latina, 2, 1–5: Göttingen, 1963–7). Continued by Paul Gerhard Schmidt (Carmina medii aevi posterioris latina, 2, 7–9: Göttingen, 1982–6).

WATKINS, CALVERT, 'Indo-European Metrics and Archaic Irish Verse', *Celtica*, 6 (1963), 194–249.

ZIOLKOWSKI, JAN M., 'The Nature of Prophecy in Geoffrey of Monmouth's *Vita Merlini*', forthcoming in the proceedings of a 1986 Harvard-Yale 'Poetry and Prophecy' conference.

'The *Occupatio* by Odo of Cluny: A Poetic Manifesto of Monasticism in the 10th Century', *Mittellateinisches Jahrbuch*, 24–25 (1989–90), forthcoming.

3 The Quotation in Goliardic Poetry: The Feast of Fools and the Goliardic Strophe *cum auctoritate*

PAUL GERHARD SCHMIDT

Despite their apparent lightness and spontaneity, the secular lyrics written by the clergy in the twelfth and thirteenth centuries cannot be read and understood without assistance. Their many borrowings from, allusions to, and reminiscences of the Bible, liturgical texts, ancient and medieval authors, need to be identified.

In recent decades research has paid attention to quotations and parodistic *contrafacta*. This has at last produced critical editions—some of them with commentaries—for the Archpoet, for Walter of Châtillon, and for that great composite manuscript the Codex Buranus. The artistic character of goliardic poetry, rich in generic conventions, has in consequence been illuminated more clearly. In some cases it has also proved possible to identify the authors of poems that have been transmitted anonymously. They emerge

The following remarks, delivered as lectures on different occasions from 1970 to 1972, were published in *Antike und Abendland*, 20 (1974), 74–87, under the title 'Das Zitat in der Vagantendichtung: Bakelfest und Vagantenstrophe *cum auctoritate*'. Thanks are due to de Gruyter Verlag for permission to republish this augmented version and to Peter Godman for the English translation.

The essay is frequently cited, and its thesis has been generally accepted, most recently in Udo Kindermann's lecture to the third meeting of the 'Mediävistenverband' in Paderborn, March 1989, entitled 'A la feste sui venuz et ostendam quare: Ein Gegenfest schafft lateinische Literatur' (corroborative rather than innovatory). Additional bibliography: W. Arlt, *Ein Festoffizium des Mittelalters aus Beauvais in seiner liturgischen und musikalischen Bedeutung* (Cologne, 1970), i. 38–51; W.-D. Lange, 'Stilmanier und Parodie: Zum Wandel der mehrsprachigen Dichtung des Mittelalters', in *Literatur und Sprache im europäischen Mittelalter: Festschrift für K. Langosch* (Darmstadt, 1973), 398–416; F. J. Worstbrock, 'Rhetorische Formtypen der mittelalterlichen Lyrik', *Deutsche Vierteljahrsschrift für Literaturwissenschaft und Geistesgeschichte*, 49 (1975), 8–31 (esp. 19–21); J. Mann, 'Satiric Subject and Satiric Object in Goliardic Literature', *Mittellateinisches Jahrbuch*, 15 (1980), 63–86 (esp. 79–82); G. W. G. Wickham, 'Robert Grosseteste and the Feast of Fools', *Sewanee Mediaeval Colloquium Occasional Papers*, 2 (1985), 81–99.

not as *clerici vagantes*, but as holders of high ecclesiastical office, as officials in a royal chancery, or as respected masters in the schools. To only a small number of drinking and begging songs can the long-entrenched concept of goliardic poetry be applied in the more restrictive sense of the term. The greater part of the poetry which goes by that name is poetry of the schools and of the clergy, and in it the learned element is dominant.[1]

How decisively the content and structure of entire poems are influenced by individual literary models was not generally recognized by the first editors of these texts. Hence the long-standing failure to see that Walter of Châtillon's 18th poem represents a paraphrase of Psalm 50 or that the anonymous lover of the composite manuscript of Ripoll created one of his love poems (no. 23) by basing it closely on Ovid, *Amores* 1. 5.[2] Both of these medieval *remaniements* can serve as model examples of the great differences in the extent of verbal borrowings. The anonymous love-poet, who differs from Ovid in metre, content, and number of verses only to a slight extent, none the less consciously avoided following his model word for word. His restraint in not employing formulae and characteristic expressions from Ovid cannot be said to alter the fact that, in his works, we find one of the closest imitations of Ovid's erotic poetry. Walter of Châtillon, by contrast, opens almost every strophe of his paraphrase of the Psalms with a verbatim quotation of the relevant psalm-verse. To be sure, he never signals his source in any of the many literal quotations; no outward peculiarity distinguishes the borrowings from the otherwise biblically-coloured language of his poem. This practice of employing quotations without indicating or referring to their authors can be observed not only in the example I have just cited but also more generally in the secular poetry written by clerics.

As a rule the quotations were inserted unobtrusively into the new context, and in the process alterations to their original form were frequently unavoidable. In this manner a *metrical* quotation, for example, can be adapted to a *rhythmical* context. In his great

[1] H. Schüppert, *Kirchenkritik in der lateinischen Lyrik des 12. und 13. Jahrhunderts* (*Medium Aevum* Philologische Studien, 23: Munich, 1972), 29–32.

[2] K. Strecker, *Moralisch-satirische Gedichte Walters von Châtillon* (Heidelberg, 1929), 152; L. N. d'Olwer, 'L'escola poetica de Ripoll en els segles X–XIII', *Institut d'Estudis Catalans*, 6 (1923), 41 f. Cf. also W. Offermanns, *Die Wirkung Ovids auf die literarische Sprache der lateinischen Liebesdichtung des 11. und 12. Jahrhunderts* (Beihefte zum *Mittellateinischem Jahrbuch*, 4 [3]: Wuppertal, 1970), 91 f. The Ripoll collection is edited by Th. Latzke, *Mittellateinisches Jahrbuch*, 10 (1975), 138–201 (esp. 175 f.).

complaint-poem *De superfluitate clericorum* the satirist Gillebert refashioned an hexameter of Juvenal's into a goliardic line—no other external characteristic signals to the reader the fact that here he is dealing with a quotation.[3]

A poem transmitted in the Codex Buranus (no. 15), which makes fairly extensive use of quotations, also provides no indication of the borrowings which appear regularly at the beginning of each strophe. In addition the poet also ends the final strophe with a quotation; he adduces at the conclusion the same Horatian expression (*Epistles* 1. 11. 27) with which he had opened the first strophe. *Caelum, non animum* is the theme and substance of this poem, in which no reference is provided for the quotations, despite their importance in its structure.[4] A counter-example is presented by *Carmen Buranum* 1, the author of which expressly emphasizes that he is citing a binding pronouncement: *Date, vobis dabitur; talis est auctoritas*. This *auctoritas*-quotation from Luke, also used in two other goliardic poems, cannot be overlooked as the many hidden borrowings from the heritage of tradition in the language of secular poetry written by clerics have been in the past.[5]

Here I do not intend to point out quotations which have been overlooked, but I should like to draw attention to a noteworthy style of citation which provides the constitutive principle according to which a series of poems is structured. Here, in almost every case, an *auctoritas* appears at the end of each strophe; frequently it is explicitly designated a quotation in the preceding line or announced by a reference to its author. Mostly one is dealing with a hexameter or pentameter taken more or less verbatim from an

[3] Gillebert, *De superfluitate clericorum*, ed. R. B. C. Huygens, *Sacris Erudiri*, 13 (1962), 519 ff.:

1 Ad scribendum equidem ebes et piger sum,
sed cum vulgus videam viciis immersum
clerumque conspiciam undique perversum,
si natura denegat, facit ira versum.
(Cf. Si natura negat, facit indignatio versum: Juvenal 1. 79)

[4] O. Schumann, in his commentary on *CB* 36 (*Carmina Burana*, ii. 1 (Heidelberg, 1930), 56–9) points to a total of four poems in the Codex Buranus, in which the initia are repeated at the end: 14, 15, 35, and 36. In *CB* 14 and 36 some of the strophes also begin with quotations (Walther, *Proverbia* 33702, should supplement *CB* 36, 2a).

[5] *CB* 1. 5. 1 ff. after Luke 6: 38. Cf. also *CB* 42. 7. 3 and 218. 2. 3. For previously unrecognized quotations and borrowings in the Archpoet see W. Heckenbach, *Mittellateinisches Jahrbuch*, 4 (1967), 145–54 and O. Zwierlein, ibid. 7 (1972), 102–21. Cf. P. Dronke, below, pp. 57–71.

ancient poem and attached to a strophe by a disyllabic end-rhyme that it shares with three rhythmical lines. Its unaltered metrical form clearly distinguishes it from its rhythmical context. The distinctive character of the borrowing is thereby clearly preserved. An example is provided by a goliardic strophe of Walter of Châtillon which uses the same quotation from Juvenal that Gillebert had refashioned into a thirteen-syllable; Walter, by contrast, did not alter Juvenal's hexameter.[6] His strophe can serve as a model for the regularly formed 'goliardic strophe *cum auctoritate*'. This term, to which Wilhelm Meyer gave currency, frequently appears in most of the relevant literary histories and introductions to medieval Latin versification, together with the supposition that Walter of Châtillon is to be regarded as the originator of this manner of quotation.[7] There has, as yet, been no study of the form, its function, or its purpose. In this essay I shall attempt to provide a survey of the *auctoritas*-poems which are scattered in out-of-the-way editions and to offer an explanation for the genesis of their form. More than forty poems from the period between 1160 and 1300 are distinguished by this structure—among them complaint- and debate-poems, fables, encomia and victory songs, invectives and visions. The versified literary history by Hugh of Trimberg, the *Registrum multorum auctorum*, also, in some of its shorter passages, concludes the strophes with a metrical quotation borrowed verbatim.[8]

It seems to me that the development of *auctoritas*-poetry was completed in two phases; in a modified form it was employed beyond the bounds of the genre. Among the first examples are

[6] Walter of Châtillon, *Moralisch-satirische Gedichte*, ed. K. Strecker (Heidelberg, 1929):

5.4 Cum mundum intuear sordis fluxu mersum
et nature penitus ordinem perversum
et hunc a principibus in vulgus dispersum:
Si natura negat, facit indignatio versum. (= Juvenal 1. 79).

[7] O. Hubatsch, *Die lateinischen Vagantenlieder des Mittelalters* (Görlitz, 1870), 72; W. Meyer, *Gesammelte Abhandlungen zur mittellateinischen Rythmik*, 3 vols. (Berlin, 1905–36); repr. Hildesheim, 1970), i. 304, 309, 333, iii. 356 ff.; K. Strecker and R. B. Palmer, *Introduction to Medieval Latin*[2] (Berlin, 1963), 84 f.; A. Önnerfors, *Mittellateinisches Jahrbuch*, 3 (1966), 72; P. Klopsch, *Einführung in die mittellateinische Verslehre* (Darmstadt, 1972), 33. On the term *auctoritas*: K.-H. Luetcke, '*Auctoritas* bei Augustin', *Tübinger Beiträge zur Altertumswissenschaft*, 44 (Stuttgart, 1968); M.-M. Davy, 'Les *auctoritates* et les procédés de citation dans la prédication médiévale', *Revue d'histoire franciscaine*, 8 (1931), 344–54; W. Hessler, 'Auctoritas im deutschen Mittellatein', *Archiv für Kulturgeschichte*, 47 (1965), 255–65.

[8] Cf. A. Vernet, 'Poésies latines des XIIe et XIIIe siècles (Auxerre 243)', *Mélanges F. Grat* (Paris, 1949) ii. 267–74; K. Langosch, *Das 'Registrum multorum auctorum' des Hugo von Trimberg* (Germanische Studien, 235: Berlin, 1942) 91 ff.

several of the moral-satirical poems of Walter of Châtillon. In them an announcement often precedes the quotations. Before a verse from Ovid, for example, it is *quia iuxta nobilem versificatorem*; Horace—on account of the rhyme—is designated *lascivus*.[9] In poems of the second phase, announcements of this kind are seldom found; in quite a number of cases they are wholly missing. In yet another respect Walter is distinguished from the second phase. He seems to use the new form predominantly as a vehicle for a specific theme. His poems are songs of complaint in which the higher clergy is reproached with its failings; attacks are directed by preference against a vice of the clergy, its *avaritia*. At the same time Walter laments the penurious state of the students and admonishes the clerical holders of offices, whom he also disparages, to perform their duty of *largitas*. This unique combination of abuse and entreaty is linked to an external occasion.[10] Walter defines it in two *auctoritas*-poems: it is the Feast of Fools. This label covers a series of festivals which Wilhelm Meyer once characterized as the 'great Dionysia of the Catholic Church'.[11] On the feast days of Stephen, John, the Massacre of the Holy Innocents, and the Circumcision the lower clergy in the cathedral churches assumed the functions which were otherwise reserved for their superiors in the hierarchy. For example, a boy-bishop, elected from the ranks of the younger clergy, made his appearance, bearing the sign of episcopal office, the *baculus*, and took the lead in the ecclesiastical celebrations of the day, in which numerous

[9] Walter of Châtillon:

6.17 Semper habet comitem paupertas merorem,
perdit fructum Veneris et amoris florem,
quia iuxta nobilem versificatorem
non habet unde suum paupertas pascat amorem. (= Ovid, *Rem.* 749)

4.13 A prelatis defluunt vitiorum rivi,
et tantum pauperibus irascuntur divi;
impletur versiculus illius lascivi:
quicquid delirant reges, plectuntur Achivi. (= Horace, *Ep.* I. 2. 14)

[10] Praise and flattery on the one hand, insult and calumny on the other alternate erratically. The editor of Walter's poems often pointed to these abrupt changes of direction (Strecker, 61 ff., 72 ff., 123) and regarded them (mistakenly, in my view) as a sign of disruption in the transmission.

[11] Cf. H. Spanke, 'Zu den Gedichten Walters von Châtillon', *Volkstum und Kultur der Romanen: Sprache, Dichtung, Sitte*, iv (Hamburg, 1931), 197–220 (204). On the Feast of Fools cf. G. M. Dreves, 'Zur Geschichte der fête des fous', *Stimmen aus Maria Laach*, 47 (1894), 571–87; E. K. Chambers, *The Mediaeval Stage* (Oxford, 1903), i. 274–335; H. Spanke in *Studi Medievali*, NS 8 (1935), 50 f.; B. Bischoff, 'Vagantenlieder aus der Vaticana', *Zeitschrift f. rom. Phil.* 50 (1930), 76–95; Schüppert (n. 1 above), 35 ff., 54 ff.; Ch. Witke, *Latin Satire: The Structure of Persuasion* (Leiden, 1970), 241 ff., 255 ff.

extensions to the liturgical text were evinced. Thus the choir performed solemn, hymn-like songs in praise of the *baculus* and its bearer. Furthermore, hierarchical order was broken down and abolished by masked antics, colourful pageants, boisterous dances, and theatrical performances. The turbulent activities ended with the *Festum asinorum* or *Festum stultorum*, the festival of the subdeacons, that is, the schoolboys and students. On that day it was the custom to attack the higher bearers of ecclesiastical office in mocking songs, under the protection of the *libertas Decembriae*. Because of their excesses these festivals caused mounting irritation; the history of the Feast of Fools is in part the history of its prohibitions. Regrettably we lack precise details about the organization of the celebrations; thus we know but little about who supplied the considerable financial support required for the Feast of Fools. Certain witnesses speak of the *baculifer* being obliged to provide a meal for the participants in the festival, which would have far out-taxed his financial resources in the circumstances. From the letters of Guido of Bazoches we learn that he was the bearer of the *baculus* as a young man in Châlons and took his responsibilities as a fool so seriously in the course of it that, as a result of the debts he had contracted during the Feast, he was obliged to avoid his native city for a long time.[12]

The financial burden of the festival was, as this single example shows, not to be borne alone by the youth in charge of the celebrations. The festival certainly did not remain confined to a narrow circle of the *clericuli* who belonged to the cathedral but attracted many guests and strangers, among them travelling singers and poets who possibly acted in formal poetic competitions and are bound not to have moved on from them without some reward.[13] On the evidence of the appeals to the *largitas* of the prelates—seldom absent from poems performed at the Feast of Fools—I suspect that a large part of the expenditure necessary to ensure the success of the Festival was contributed by the holders of benefices who were abused on that day. This may explain the confusing juxtaposition of mockery and entreaty. The poet who wrote a song for the Feast of Fools consequently faced the difficult task, requiring a good deal of tact, of so gauging the pillory

[12] *Liber Epistularum Guidonis de Basochis*, ed. H. Adolfsson (Studia Latina Stockholmiensia, 18: Stockholm, 1969), nos. 18 and 23 (89. 9 ff.: Bacularis illa festivitas seu baculus ille festivus . . . fuit mihi poculi venenum sub melle species occultantis); W. Wattenbach, *Neues Archiv*, 16 (1891), 69 ff.

[13] Bischoff (n. 11 above), 88; Spanke, *Volkstum* (n. 11 above), 212 ff.

traditional in the *Festum Stultorum* that he gave equal satisfaction to the young men and to the holders of ecclesiastical office. When the poet appeared as the mouthpiece of recognized authorities, when he aroused in his audience expectations of a quotation by the regular use, at the end of each strophe, of citations (which could even be guessed, in the circumstances, on the basis of the end-rhyme in the preceding three lines)—this new dramatic element which he introduced into his poem offered him the opportunity of creating a consensus with his heterogeneous public. The receptiveness, ubiquitous in the Middle Ages, to parodies—a taste shared by the entire learned public—favoured this virtuoso play with quotations. I should like to adduce only one example of the parodic undertones in the *auctoritas*-poems: the frequently-employed formula of fulfilment which declares that an event long-prophesied has now come about. Sometimes it appears in the phrase 'impletur versiculus'; in other places in the form 'Ergo sic impletum est, quod dicunt auctores'.[14] Beyond any doubt the model is to be found in sacred language. In the liturgy of the Christmas festival, immediately followed by the days of the Feast of Fools, the Old Testament prophecies and the Gospel-accounts are intermingled with one another. The Feast of Fools makes use of the formula of fulfilment in a style suited to its character: just as it allows a boy-bishop to act as *episcopus*, so it raises the classical school authors to a rank comparable to that of the Psalmists and the Prophets whose place they take.

Maxims from ancient poetry and Old Testament prophecies are casually adduced, one beside the other, in the same poem as *auctoritates*; but the biblical quotations, frequently refashioned into goliardic lines, appear only in poems of the first phase, whose texts often display a paraliturgical character.[15]

One may perhaps trace the curious outward form of the quotation-poems back to liturgical models. An explanation for the form which Walter chose for his satirical poems has recently been

[14] Walter of Châtillon 4. 13. 3 (cf. above, n. 9); 6. 18. 3; 7. 12. 3 (*completum*). Other parallels can be adduced thanks to the concordance of Violetta de Angelis, *Gautier di Châtillon, Poesia lirica: Concordanza* (Pavia, 1988).

[15] I employ the term *auctoritas* more broadly than Strecker, who clearly used it only to refer to quotations from classical poetry. Consequently he often remarks, in his notes to Walter's poetry, that some of their strophes end without an *auctoritas* (62, 73). None the less, quotations from the Bible are often used instead, quotations which differ only in respect of form from the other *auctoritates*. I therefore count Walter's 8th poem, with its numerous biblical quotations at the end of the strophe, among the *auctoritas*-poems and therefore among poems written for the Feast of Fools.

sought in the polymetry of Menippean satire;[16] I should rather point to the tropes, expansions on the liturgy in which are contained the seeds of medieval religious drama. Since the eleventh century they had contained individual hexameters intercalated into a rhythmical context.[17]

Against the background of the Feast of Fools so many distinctive features of the quotation-poems are explained that I have no doubt that this genre owes its origins to the occasion of the festival. The link which I have posited between *auctoritas*-poetry and the *Festum Stultorum* is manifest in both of Walter's poems which expressly name the festival. Four more quotation-poems by Walter, which similarly contain invectives against the higher clergy and admonitions to *largitas*, can be understood as texts to be recited at the Feast of Fools. They have in part already been claimed for this festival by Hans Spanke, on the basis of internal criteria.[18] Only Walter's first poem seems to contradict my thesis. Although, from the formal point of view, it belongs to the group of *auctoritas*-poems, it has no obvious connection with the festival.[19] Indeed, Karl Strecker in his edition described it as a request to the Pope for a benefice. Because he took the view that the poet had recited it personally to the Pope, he concluded from it that Walter had stayed in Rome. In taking this view, he followed the hint of younger manuscripts in which the poem had been construed in the same manner and provided it with suitable headings; in Codex B it reads, for example, *coram domino papa in consistorio.* Doubts about this interpretation were first voiced by Spanke soon after the appearance of Strecker's edition. Spanke, however, chose not to discuss the problematic poem in detail and confined himself to the suspicion that it could be a fiction or a stylistic exercise.[20]

I believe that the riddle of this poem can be resolved by taking its character as a quotation-poem into account. Walter begins by praising a person no more precisely defined than by the terms *vir tantus* (1. 1), *iudex* (2. 6; 4. 1) or *pastor bonus* (which he uses

[16] Witke (n. 11 above), 261 f.

[17] P. Evans, *The Early Trope Repertory of Saint Martial de Limoges* (Princeton Studies in Music, 2: Princeton, 1970), 64 f.

[18] The Feast of Fools is mentioned in 4. 3. 1 and 13. 4. 2; 5, 6, 7a, 8, and 12 were in my opinion also intended for this festival.

[19] Instead of the 'Vagantenstrophe' the *Stabat-Mater* strophe is used here; the strophes conclude with a seven-syllabic line. The quotations from classical poetry (in comparison to the smaller number of biblical quotations) are correspondingly shorter. Instead of a complete hexameter only a hemipes is employed (e.g. in the second strophe, Ovid, *Rem.* 2 *bella mihi video*; in the third strophe, Ovid, *Rem.* 4 *te duce signa tuli*).

[20] *Volkstum* (n. 11 above), 214.

intermittently), and whose favour he entreats. Protected by him, Walter dares to complain about the depravity of the world. Above all he criticizes the greed of the prelates and seeks to arouse sympathy for the penurious condition of the students. Their burdensome study of the *artes*, and reading of '*Arma virum*', '*Fraternas ascies*', and '*Bella per Emathios*' bring them no profit; the student of the *artes* and the theology student gnaw at the bone of hunger (17. 6). The poet provides detailed examples of the study of theology with numerous instances of the allegorical interpretation of scripture. He begins with the *virga Jesse* frequently mentioned at Christmas and the fleece of Gideon signifying the Virgin Birth; among the examples that follow, those in which a staff—such as that of Moses—is interpreted as a prefiguration of the Cross, are dominant. The learned statements about staffs mentioned in the Bible end with a request, directed to the poet's patron, for a prebend or another gift which would enable him to continue his studies free from external duress.

The use of this poem as a song for the Feast of Fools, which, in my opinion, is revealed by its form, can be supported at many points by arguments based on its contents: to the Feast of Fools belong the laments on the *avaritia* of the severely criticized prelates, the depiction of the oppressed state of the students, the appeal to *largitas*, and also the mention of the *virga Jesse* which plays a special role in the liturgy and lyrics of Christmas as well as in the songs for the Feast of Fools.[21] The collection of examples of significant staffs can readily be interpreted as the customary praise (here, it must be admitted, obliquely introduced) of the *baculus* and thereby of its bearer. Only the person whom the poet asks for a benefice does not fit into the setting of the Feast of Fools so long as he is identified with the Pope. This identification is, to be sure, suggested by an expression of Walter's—*ad Romani sedem patris . . . sum reversus*.[22] None the less, I do not think that this proves an insuperable obstacle to an interpretation that differs from Strecker's. The presumed patron whom the poet extols so euphorically need not be the Pope: it is quite enough to think of a person who represented the *summus pontifex*. We have seen that masquerades and exchanges of roles in the exercise of ecclesiastical office were typical forms of the Feast of Fools. What prevents us from making the assumption that Walter, on this occasion, did not recite his quotation-poem before a boy-bishop, but before a *baculifer* who

[21] Ibid. 210. [22] Walter of Châtillon 1. 26. 4 ff.

appeared in the role of the holder of the highest ecclesiastical office? Moreover we encounter a *papa scholasticus* some years before Walter's poem in an exuberant begging-song by Abelard's pupil Hilarius, which was also recited at a school festival.[23] If Walter later apostrophizes the Pope of his poem twice as *iudex*, this manner of address is easily explained by the fact that, during the feast, it was indeed the *baculifer* who awarded the title of victor in the competition between poets. Finally I should like to adduce a further argument from the manuscript-transmission of Walter's poem to support my view that Walter's first poem was intended for performance at the Feast of Fools. A manuscript (Db) contains a version, the additional strophes of which plainly reveal that the poem was recited in this revised form at a Feast of Fools. In three new concluding strophes the clergy of Troyes are admonished, *inter alia*, to exclude the *leccatores*, the lickers of plates, from the feast. Perhaps the passing shot is directed at irksome rivals of the poet; this I would very much suspect in the context of a poetic competition.

Reworkings of poems by Walter are not uncommon; practically every manuscript reveals traces of such alterations and interference. The fifth poem, also a song for the Feast of Fools, was in all probability recited in many forms and certainly in various places.[24] In three strophes which have been added later a certain Fulmarus is explicitly excepted from the invective against the avaricious. The poet hopes for no small reward for his praise of Fulmarus. He, who had come to the Feast of Fools from afar, did not want to have to return on foot to his homeland. His request for a means of transport is couched in the hexameter 'magnus eris, si me bipedem scis quadrupedare'. As far as I know, this verse is not borrowed from ancient poetry. Nor can I tell whether the request for a horse derives from a medieval poem (perhaps one that has not come down to us) or was composed *ad hoc*. In any case its metrical form stands in contrast to the goliardic lines and makes the hexameter appear as an *auctoritas*. This interpolated version of Walter's fifth poem, directed to Fulmarus, is possibly one of the last quotation-poems of the first phase, which extends from 1160 until 1200. Beside Walter of Châtillon and those who rework his verse a series of anonymous poets, mostly influenced by Walter,

[23] *Hilarii Versus et Ludi*, ed. J. N. Fuller (New York, 1929), no. XIV. 96 f., 29 f.; ed. M. L. Bulst and W. Bulst (Leiden/Cologne, 1989), 47 f.

[24] Codex B has the heading 'Treveris in capitulo'. Is the allusion at 5. 7, 1 to the Pope (as Strecker tacitly assumes) or rather to the *papa scholasticus*?

together with the author of the poems transmitted in the goliardic manuscript now in the Vatican, belong among the authors of this period. The poems of the last author, among which admittedly only one quotation-poem is to be found, were recited between 1160 and 1185 in Chartres at the Feast of Fools.[25] It must remain uncertain whether the poet of the Vatican songs or Walter of Châtillon is to be regarded as the inventor of the *auctoritas*-poems until we succeed in dating more precisely songs that belong to the same period.

The poets of the first phase intended their quotation-poems exclusively for recitation at the Feast of Fools. The quotations serve in their poems to strengthen and support their own assertions; by recourse chiefly to the Roman satirists their authors exculpate themselves from responsibility for the harsh tone of their verses. After the year 1200 we no longer find aggressive songs for the Feast of Fools; this is to be traced back to a draconian reform of the Festival which was then undertaken by Petrus de Corbolio and caused the volcanic celebrations to subside into quiescence for a space of decades.

The form of the quotation-poems had, however, already developed independently and had emancipated itself from its close links with the Feast of Fools. When the second phase began I cannot determine precisely; the earliest evidence, however, dates from the end of the twelfth century, with the result that there must have been some years when both phases evolved in parallel. At this stage the genre was practised particularly in the schools; evidence of this is provided by the numerous disputes and fables with *auctoritates*.[26] How closely quotation-poems were tied to didactic purposes is clear from a poem which deals with Ovid's works and writings in goliardic lines. 'Cum Nasonis opera nunc incipiamus' is how the author—a teacher, beyond the shadow

[25] Bischoff (n. 11 above), 78, who identified Peter of Blois as the author of these songs for the Feast of Fools. Cf. R. W. Lenzen, *Überlieferungsgeschichtliche und Verfasseruntersuchungen zur lateinischen Liebesdichtung Frankreichs im Hochmittelalter* (diss. phil. Bonn, 1973), and P. Dronke, *The Medieval Poet and his World* (Rome, 1984), 281–339.

[26] H. Walther (rev. P. G. Schmidt), *Das Streitgedicht in der lateinischen Literatur des Mittelalters* (Munich, 1920; repr. Hildesheim, 1984). Quotations are often not identified in previously published texts: L. Hervieux, *Les Fabulistes latins, 2,436–479* (*Romuleae fabulae . . . in rhythmicum sermonem medio aevo translatae*) (Paris, 1884). Cf. M. Manitius, *Geschichte der lateinischen Literatur des Mittelalters* (Munich, 1931), iii. 773. Contrary to Manitius' opinion the hexameter-*auctoritates* were not consistently composed by the poets who wrote these fables. (A. Kabell, *Metrische Studien*, ii (Acta Universitatis Upsaliensis, 6: Uppsala, 1960), 101, dates (mistakenly, in my opinion) this collection of fables to the first half of the twelfth century). Cf. further *Ysopet-Avionet*: *The Latin and French Texts*, ed. K. McKenzie and W. A. Oldfather (University of Illinois Studies in Language and Literature, 5: 1919/1921).

of a doubt—introduces his theme: the *auctoritates* of the versified *accessus* naturally originate in Ovid's poetry.[27] In other poems the figure of the teacher takes the limelight; the pupils compare their master with Apollo on account of his wisdom, with Narcissus on account of his beauty.[28] Generally this kind of flattery is only the prelude to a request for holidays; 'valde sumus fatigati' lament the pupils and remind their teacher of the saw from the *Disticha Catonis*—'interpone tuis curis gaudia'—or ask, together with the exiled Ovid, 'parte leva minima nostras et contrahe poenas'.[29]

In one case a young *magister* seems to have applied for a vacant Chair with a quotation-poem. In order to remove all doubts about his youth and lack of experience, he reports a vision to the *doctores nobiles* who are to decide as to whether he should be received into their number. The *virgo Grammatica*, he claims, has appeared to him in a vision and has encouraged him to apply with the verse from Lucan 'tolle moras, semper nocuit differre paratis' (1. 281). If the teaching office is bestowed upon him, there will be no lack of success, for does not Ovid assure us 'fac tamen incipias, sponte disertus eris' (*Ars Amatoria* 1. 610)?[30]

Let me add parenthically that the wording of this borrowing from the *Ars amatoria* enables us to identify the branch of the transmission which was drawn on by the poet. According to the testimony of most manuscripts, with which editors in general have agreed, the first half-verse reads 'fac tantum cupias'; only some of the *recentiores* offer the reading 'fac tantum incipias', which is nearest to the 'fac tamen incipias' of our *auctoritas*. Were one to succeed in dating and localizing these *recentiores* which are gathered together in the apparatus criticus under the composite siglum '*pauci*'; were one to be able to demonstrate that the variant occurs only in manuscripts of a distinct provenance, then one could draw conclusions as to the time and place at which the anonymous quotation-poem was produced.

Medieval refashionings of the shape of ancient texts can also be demonstrated in a considerable number of quotation-poems. On occasion there are adduced as *auctoritates* verses from Horace which are read only in the apparatuses of our editions, for they owe their existence to the Middle Ages: e.g. the verse 'oderunt

[27] H. Hagen, *Carmina medii aevi maximam partem inedita* (Berne, 1877; repr. Turin, 1961), 207 ff.

[28] *The Parisiana Poetria of John of Garland*, ed. T. Lawler (New Haven and London, 1974).

[29] E. du Méril, *Poésies inédites du moyen âge* (Paris, 1854), 295 ff.

[30] A. Bömer, 'Eine Vagantenliedersammlung des 14. Jahrhunderts', *ZfdA* 49 (1908), 227 ff.

peccare mali formidine poenae', which is formed on analogy with *Epistle* 1. 16. 52 ('oderunt peccare boni virtutis amore'). In a poem, *Proteus*, that borrows its *auctoritates* solely from Horace, these verses are employed after one another without interruption.[31] As the apparatus of Klingner's edition of Horace shows, the medieval addition is transmitted only in two manuscripts of the mixed class Q.

Other quotation-poems of the second phase go beyond the bounds of this school. Here one should mention several invectives against ecclesiastical orders, a *Carmen de vitiis capitalibus*, a song on the victory of the city of Parma over Frederick II in the year 1248 and a paean on the victorious struggle of the English against the Scots from the end of the thirteenth century. There are also a crusading song, estate-satires, and misogynistic poems: all in all, a colourful spectrum of disparate contents.[32] The external forms of the quotation-poems also become more various. If in the first phase the combination of the *auctoritas* with three thirteen- or three fourteen-syllable lines was the rule, the quotation is now also linked with two rhythmical lines. Instead of several goliardic lines the quotation is preceded even by the *Stabat-Mater* strophe, or other, more artistic, strophic forms. On occasion, there are departures from the consistently regular use of quotations. In *Carmen Buranum* 105 only about the last six of a total of eleven strophes end with quotations. Within poems of greater dimensions self-contained units of five to ten strophes display *auctoritates*; which is the case with the *Registrum multorum auctorum* of Hugh of Trimberg.

[31] Ed. C. Blume, *Analecta hymnica*, 33 (Leipzig, 1899), 340 ff.

14 Coronatur merito bonitatis flore,
Qui abhorret ideo pravo frui more
Nec virtutis nitido fulgeat splendore,
Oderunt peccare boni virtutis amore. (= *Ep.* 1. 16. 52)

15 Cui detur gratia operandi bene
Nunquam cesset talia perpetrare plene,
Ni laedentis verbera sentiat habenae;
Oderunt peccare mali formidine poenae. (= Ps.-Horace)

[32] F. Helfenberger, *Drei lateinische Gedichte des XIII. Jahrhunderts* (Diss. phil. Freiburg/Schweiz, 1928); J. Werner, 'Ein satirischer Rhythmus des dreizehnten Jahrhunderts', *Festgabe H. Blümner* (Zürich, 1914), 357–73; *Analecta hymnica*, 33 (Leipzig, 1899), no. 229, 221 f.; Bömer (n. 30 above), 211 ff. and 233 ff.; T. Wright, *The Political Songs of England* (Camden Society, 6: London, 1889), 106 ff.; K. Strecker, 'Walter von Châtillon und seine Schule', *ZfdA* 64 (1927), 97–125, 161–89; id., 'Quid dant artes nisi luctum!', *Studi Medievali*, NS 1 (1928), 380–91; id., 'Ein Gedicht Walters von Châtillon?', *Atti dell' Accademia degli Arcadi*, Anno XIV, NS 5/6 (1931), 47–55.

Despite this great breadth of variation some common features can also be distinguished within the second phase. Striking, first of all, is the greater care, bordering upon pedantry, with which authors ensured that the quotations remain unaltered in their wording. Walter of Châtillon, the leading exponent of the first phase, by contrast, had not flinched from making small alterations and had adapted the *modus* and *numerus* of a quotation to the requirements of his strophe. In order to acquire a suitable rhyming word at the end of a quotation he interfered with the word-order by transposition; once he shortens an hexameter of Lucan's because the verse could be adapted to the rhyme and sense he wanted only in the form of a catalectic.[33] While further on in the first phase the poets cited were repeatedly adduced by name and their verses announced by expressions designed to draw attention to them, these kinds of references occur less frequently in the second phase. But not only is there a difference between the form and presentation of the quotations in both phases, the group of authors cited also alters significantly. Verses from Horace, Juvenal, Virgil, Ovid, Lucan, Claudian, and the *Disticha Catonis* continue to be borrowed, and are frequently linked with rhyming words that had been found earlier. But in addition the circle of authorities had widened to include medieval authors. Perhaps this renewal was in response to the stimulus provided by Walter of Châtillon, who otherwise (it is true) never adduces any medieval quotations but once cites a verse from one of his own poems.[34]

The medieval quotations, previously seldom identified, in the poems of the second phase derive from the comedies *Geta* and *Pamphilus*, from the poems of the Archpoet and of Primas of Orleans, from the *Tobias* of Matthew of Vendôme, from the epic on Alexander the Great by Walter of Châtillon, from the *Anticlaudianus* of Alan of Lille, and from the *Architrenius* of John

[33] Walter of Châtillon:

5.1 Multiformis hominum fraus et iniustitia,
letalis ambitio, furtum, lenocinia
cogunt, ut sic ordiar conversus ad vitia:
quis furor, o cives, que tanta licentia!

(Lucan 1. 8, quis furor, o cives, que tanta licentia ferri?)

[34] Walter of Châtillon:

14. 2. 4 quia, sicut alibi per me declaratur,
large si non agitur, rosa derosatur.

(= Walter of Châtillon, *Lieder*², ed. K. Strecker (Berlin/Zürich, 1964), 41; 24, 3, 5 f.).

of Hauville:[36] from works, therefore, which without exception belong to the second half of the twelfth century. In the thirteenth century most of these poems found an established place in the canon of school authors; even the quotation-poems reflect in their turn the intellectual position of this 'century without Rome' in which the *Aeneis* is supplanted by the *Alexandreis*.

A further variant on the use of quotations is presented by poems that borrow their *auctoritates* from only one author, like the Ovid-*accessus* I have already mentioned, the poem entitled *Proteus* referred to above, which combines its moralistic reflections with quotations from Horace's *Epistles*, and finally the only *auctoritas*-poem transmitted in the Codex Buranus, no. 105, which again adduces only quotations from Ovid, most of them from the second book of the *Ars amatoria*.[36] In other poems the first lines of ancient and medieval poems are used by preference. With many poems of the first phase one cannot avoid the impression that their authors are solely concerned to master a form and to free themselves from a burden which they viewed as tedious. An *ars poetica* of the thirteenth century characterizes such poems as 'Rithmi . . . quibus ociosi suum gaudent ingenium experiri'.[37] Sometimes, it has to be admitted, the demonstration of *ingenium* degenerates into a *testimonium paupertatis*. An example of an author's efforts to link quotations in a meaningful way with the preceding text is provided by a poem, composed about 1300, on the battles between the English and the Scots. Its author introduces the defeat of one side with the Horatian verse *Quidquid delirant reges, plectuntur Achivi*. In the following strophe he explains the quotation:

Plebs Archiva periit ad Dunbar in bello
ubi Scoti caesi sunt Anglorum flagello.[38]

Here the special tension which in general exists between the world of this quotation and its new surroundings is no longer maintained: the quotation has become an obstacle between strophes. The author tortuously attempts to explain as best as he can quotations which do not suit the content of the song of victory. Alan of Lille once, and in an entirely different context, criticized the arbitrary selection of quotations in his image: 'auctoritas cereum habet nasum, id est, in diversum potest flecti sensum.'[39] The waxen nose

[35] 'Languida segnities hodiernos crastinat actus' (= Johannes de Hauvilla, *Architrenius*, ed. P. G. Schmidt (Munich, 1974), I. 7a and 10b).

[36] Cf. above, nn. 27 and 31.

[37] Cf. above, n. 28.

[38] Wright (n. 32 above), 166.

[39] Alan of Lille, Migne PL 210. 333.

of *auctoritas* seems to me tangible in this poem in which the evolution of quotation-poetry comes to a conclusion.

Thus, it must be admitted, is only true of secular Latin poetry which, in the course of the thirteenth century, sank to a level of artistic inconsequentiality. However, during this period the principle of regular borrowing of lines was transferred to religious poetry and there continued to flourish for a longer period. Many hymns conclude their strophes with the *incipit* of a well-known older hymn. Josef Szövérffy has pointed out some 250 hymns with this structure, which was adopted chiefly by the members of the new mendicant orders.[40] The testimonia date from the first decades of the thirteenth century and stretch into the sixteenth century. Szövérffy has voiced the suspicion that here secular *auctoritas*-poetry influenced the form of its religious counterpart. Studies of the function of these quotations in the hymns remain to be written.

The question of audience and purpose can be answered with greater probability in the case of a subsidiary genre which Szövérffy does not mention. I refer to the English-Latin carols which were sung at Christmas outside the church in the hall of a castle or a country mansion.[41] At the beginning and end of their strophes they employ individual lines from Latin hymns. The lines they cite are borrowings from a restricted group of liturgical hymns which, to judge from their text, were entrusted to individual members of the congregation. Latin borrowings and intercalary lines in English are linked with one another in the carols by rhyme and simple connections between sentences. Even a listener who did not have a command of Latin could grasp the meaning of the theological formula in them, gaining an *entrée* to liturgical poetry through its insertion into vernacular songs. Quotation poetry, in instances such as these, therefore serves the religious instruction of the laity.[42]

[40] J. Szövérffy, 'Ein Schmuckmittel der mittellateinischen Strophen: Regelmässige Zeilenentlehnung in der Hymnendichtung', *Mittellateinisches Jahrbuch*, 7 (1972), 7–40.

[41] W. O. Wehrle, *The Macaronic Hymn Tradition in Medieval English Literature* (Diss. phil. Washington, 1933); C. Brown, *English Lyrics of the XIIIth Century* (Oxford, 1932); R. L. Greene, *The Early English Carols* (Oxford, 1935); id., *A Selection of English Carols* (Oxford, 1962); H. Gneuss, *Hymnar und Hymnen im englischen Mittelalter* (Buchreihe der Anglia, 12: Tübingen, 1968), 213–21.

[42] A collection and study of the German-Latin macaronic poems—the best-known must be *In dulci iubilo*—is still to be completed. Still indispensable are the works of Heinrich Hoffmann von Fallersleben, '*In dulci iubilo, Nun singet und seid froh: Ein Beitrag zur Geschichte der deutschen Poesie*[2] (Hannover, 1861), and Johannes Bolte, 'In dulci iubilo: Ein

In the secular poetry of the fourteenth and fifteenth centuries quotation-poems, as a form, are still employed intermittently. They appear chiefly in bilingual macaronic poetry, in which Latin *auctoritates* are linked with English, Provençal, French or multilingual lines: a symptom of the retreat of Latin and the rise of the vernaculars. None the less a drinking-song (presumably of the thirteenth or fourteenth century), with its conventional triad of wine, women, and gambling, and an *auctoritas* from the *Disticha Catonis*, is directed much more at an audience that knows Latin, an audience which is to be sought in the milieu of the schools or of the universities, membership of which must have generally corresponded with that of the Feasts of Fools in the twelfth century.[43]

The principle of regular line-borrowing which developed at the Feasts of Fools and which in its narrowest form only employs opening-lines as quotations is finally copied, in my view, in purely vernacular poetry. There seems to me a distant echo of the poems of Walter of Châtillon when Petrarch, in his seventh *Canzone*, ends each strophe with a quotation linked by rhyme to the preceding line. His strophes conclude with the beginning of a Provençal song and with the first lines of well-known Italian songs of the *dolce stil nuovo*, among which are to be found the beginnings of songs by Guido Cavalcanti and Dante. Petrarch ends the last strophe with a quotation from himself, with the first line of his first *Canzone*.[44]

Jubiläumsbeitrag aus der Geschichte der lateinisch-deutschen Mischpoesie', *Festgabe an Karl Weinhold* (Leipzig, 1896), 91–129.

[43] *Des Fames des dez et de la taverne*, ed. Barbazan, *Fabliaux et contes des poètes françois* (Paris, 1808), iv. 485 ff.:

11 Por boire le bon vin fui generatus,
Por amasser tresor non extiti natus,
Par sanblanc ne suit pas mout locupletatus,
Car avers hom en ciex jà n'iert exaltatus:
Despice divitias si vis animo esse beatus. (= *Dist. Catonis* 4. 1. 1)

Cf. O. Müller, *Das lateinische Einschiebsel in der französischen Literatur des Mittelalters* (Diss. phil. Zürich, 1919), 89 f.

[44] F. Gennrich, *Die Kontrafaktur im Liedschaffen des Mittelalters* (Summa Musicae Medii Aevi, 12: Frankfurt, 1965), 41, 172–4, points to a similar use of quotations in Provençal poetry.

4 The Archpoet and the Classics

PETER DRONKE

In a collaborative volume, *The Interpretation of Medieval Lyric Poetry*, published in 1980, I tried to analyse 'The Art of theArchpoet' by way of a close reading of one of his compositions: *Lingua balbus, hebes ingenio*.[1] The opening phrase, *Lingua balbus*, the Archpoet adapted from Isaiah, and he paired it with another, *hebes ingenio*, from Cicero.[2] This pairing is, one might say, emblematic for his poem, which unites the fervour of the Old Testament *vates* with the persuasive craft of the Roman master of eloquence. In my essay I mentioned only briefly that one of the Archpoet's characteristic virtuosities lay in his knowing alliance of classical and biblical language. He habitually combines expressions from both, playing them off against each other, with a keen sense of the connotations of the original contexts and a keen expectation that some of the subtlest, most literate members of his first audience—the court circle of Emperor Frederick Barbarossa—would share his contextual awareness and delight in it.

The year 1980, however, also saw the appearance of a very different appraisal of this poet. Francis Cairns, discussing the Archpoet's 'Confession', *Estuans intrinsecus*, claimed that its author had no real familiarity either with the classics or the Bible. According to Cairns, 'even a casual inspection reveals that many of [his] reminiscences are fleeting or trivial and . . . almost all seem to be lacking in contextual associations'.[3] The Archpoet could have got 'half remembered classical . . . tags' from grammars or florilegia, and biblical ones from the liturgy.[4] The real sources of

[1] P. Dronke, 'The Art of the Archpoet: A Reading of "Lingua balbus"', in *The Interpretation of Medieval Lyric Poetry*, ed. W. T. H. Jackson (New York-London, 1980), 22–43.

[2] Cf. *Die Gedichte des Archipoeta*, ed. H. Watenphul and H. Krefeld (Heidelberg, 1958), 87.

[3] F. Cairns, 'The Archpoet's Confession: Sources, Interpretation and Historical Context', *Mittellateinisches Jahrbuch*, 15 (1980), 87–103, at 88.

[4] Loc. cit.

the Archpoet's Confession, Cairns claimed, lay not in the classics or the Bible, but in some Cistercian treatises on penance of the poet's own day, which he was parodying. 'In comparison, the alleged Classical and Vulgate "sources" pale into insigificance.'[5]

Three years later a German scholar, Johannes Hamacher, re-examined these claims and showed tellingly that Cairns's inspection had indeed been casual.[6] The parallels between the Archpoet and Cistercian writings were far too vague and tenuous to be significant or to warrant any inference about direct relationships. By contrast, Hamacher demonstrated in detail with three biblical texts and one classical one that the full import of the Archpoet's verses does depend on full alertness to the original contexts. His classical example was Ovid's 'confession of amorousness' in *Amores* 2. 4.[7] Where Cairns had noticed 'only one point' at which Ovid's elegy is related to the medieval poem, and thought that 'we have here another case where the Archpoet is employing a tag he has met out of context',[8] Hamacher showed how pervasively this elegy was present in the Archpoet's imagination, and how it helped to shape several key aspects of the poem *Estuans intrinsecus*.

Besides Cairns's failure to recognize the imaginatively vital parallels to the Archpoet's language and thought, and to distinguish these from pointless ones and forced ones, I would also signal a historical misjudgement that underlay his approach. The *viri docti*, as the Archpoet himself calls them,[9] who were his immediate audience, were the worldly, sophisticated administrators, diplomats, and lawyers in the service of the Emperor—above all, the wily and quite unspiritual Chancellor, Reinald of Dassel, who was the Archpoet's Maecenas. Intellectually such men were a world away from the Cistercians and their penitential meditations. For a parody of such meditations even to be possible, these would have had to be familiar to Reinald and his circle. Yet historically the notion that Barbarossa's Chancellor read Cistercian works on penitence is about as likely as that Mrs Thatcher's Chancellor reads them.

[5] Ibid., Cairns, 95.

[6] J. Hamacher, 'Die "Vagantenbeichte" und ihre Quellen', *Mittellateinisches Jahrbuch*, 18 (1983), 160–7.

[7] The crucial importance of this elegy for the Archpoet's intentions had also been noted briefly in the résumé of a lecture by P. G. Walsh, 'The Archpoet's Confession', *Proceedings of the Classical Association*, 72 (1975), 18 f.

[8] Cairns (n. 3 above), 96.

[9] *Gedichte*, poem 1, st. 1 (p. 47). Citations of the Archpoet below are numbered according to this edition.

After Hamacher's essay it seems unnecessary to dwell further on the range and depth of the Archpoet's classical and biblical culture in general terms. It is time to define the specific role of classical elements in the fabric of his verse, by examining what happens in a few revealing passages.

Classical parallels to the Archpoet's verses have been noted in several editions, most extensively in that of Watenphul and Krefeld. In the very selective considerations that follow, I shall only now and then add suggestions of parallels that are new as such: I shall concentrate rather on eliciting the poetic implications of particular echoes. On the basis of this, in conclusion, I should like to hazard a guess about what the nature of these echoes may be able to tell us about the Archpoet's classical formation. What was his career before we encounter him and his poetry in the imperial court circle in the 1160s? How did he acquire his intimacy with classical poetry? How did he come to write as he did? These are questions that have scarcely yet been broached.

I shall confine my illustrations to a few strophes in the renowned *Estuans intrinsecus*.[10] The precise force of certain parallels is not always seen easily or at once. Those that may at first seem insignificant, or explainable merely as mild coincidence, may also reveal themselves as more when they are confirmed by others from the same classical poem and context, or when a fuller awareness of that context shows its relevance to diverse moments in the pattern the Archpoet was creating. Thus several scholars have seen how passages in the Book of Job provide, as it were, a *basso continuo* for the poet's own more despondent self-avowals. On the other hand, even Hamacher appears to agree with Cairns about the 'obvious superficiality of the Archpoet's contact with Horace' ('die offensichtliche Oberflächlichkeit der Berührung mit Horaz').[11] Let us reconsider this, beginning with the second strophe in the Confession:

Cum sit enim proprium viro sapienti
supra petram ponere sedem fundamenti,
stultus ego comparor fluvio labenti,
sub eodem aere nunquam permanenti.

Though it behoves a wise man
to set his 'foundation' on a rock,
I, foolish, am like the gliding stream,
never remaining under the same sky.

[10] *Gedichte*, poem 10. [11] Hamacher (n. 6 above), 165.

In the first two verses the poet blends a phrase from Matthew with one from Luke. Both, in their original contexts, designate the man who listens to Christ's words: in Matthew, he is 'like the wise man who has built his house on a rock'; in Luke, 'he has set his foundations on a rock'.[12] Both evangelists have the words *super petram*, but the Archpoet has adapted *viro sapienti* from Matthew and *posuit fundamenta* from Luke. Yet he has made a small but mischievous change to Luke's expression: it is no longer a matter of setting foundations (*fundamenta*) on the rock, but of setting on it 'the seat of the fundament' (*sedem fundamenti*).[13] The wise man, in the Archpoet's sardonic variant, is the one with *Sitzfleisch*. He himself, he now assures us, is very different. The difference is expressed in words that contain a reminiscence both of Job and Horace.

That of Job is glimpsed in the fourth verse: Job's *homo*, who 'never remains in the same condition' ('numquam in eodem statu permanet'),[14] is the model for this particular man, the Archpoet, *sub eodem aere nunquam permanenti*. The image of the gliding stream, on the other hand, may at first glance seem only tenuously related to Horace's *Epistle* 1. 2. 42 ff., where

> rusticus exspectat, dum defluat amnis; at ille
> labitur et labetur in omne volubilis aevum.
>
> the churl waits for the stream to flow away; yet it
> glides and will glide, rolling on for ever.

However, in the fifth strophe of the Confession there is an echo of the same Horatian passage, a dozen lines earlier in the same *Epistle*. Where the Archpoet declares himself

> voluptatis avidus magis quam salutis—
> mortuus in anima, curam gero cutis

[12] Matth. 7; 24: 'Omnis ergo qui audit verba mea haec et facit ea / adsimilabitur viro sapienti qui aedificavit domum suam super petram.' Luc. 6: 48: 'Similis est homini aedificanti domum / qui fodit in altum et posuit fundamenta super petram'.

Bibical citations here and below are from *Biblia sacra iuxta vulgatam versionem*, ed. R. Weber OSB (2 vols., 2nd edn., Stuttgart, 1975).

[13] While *fundamentum* in its anatomical sense, 'fundament', is well attested in Old French texts from the twelfth century (cf. Tobler-Lommatzsch, *Altfranzösisches Wörterbuch*, s.v. *fondement*), it is not recorded in *Thesaurus linguae latinae*, s.v. *fundamentum*. However, it is found in medieval Latin medical texts, the first recorded instance being in Roger of Salerno—who will have been at the medical school when the Archpoet studied there (cf. poem 6)—*Chirurgia* (ed. K. Sudhoff, *Beitr. z. Gesch. d. Chirurgie im Mittelalter* ii (1918), 3, 703. I am indebted to the scholars at the *Mittellateinisches Wörterbuch* for this information.

[14] Iob 14: 1–2.

avid of sensual bliss more than of salvation—
dead in soul, I keep my skin sleek,

in Horace it is Penelope's suitors who are 'concerned more than is right with keeping their skin sleek' ('in cute curanda plus aequo operata')—more than is right, that is, in human terms: here heavenly salvation and the soul's death do not enter in. (In another poem the Archpoet keeps still closer to this Horatian wording, as he says of Barbarossa 'His mind is not bent on keeping the skin sleek'—'Non est eius animus in curanda cute'.[15])

It is the context in Horace which makes clear that these were no isolated tags the Archpoet had picked up. Horace was making a contrast between two ways of life: 'We are mere ciphers . . . keeping our skin sleek more than is right . . . Will you not wake up, to save your life? . . . If you don't bend your mind to honourable pursuits, you'll be tormented by envy or by love . . . Dare to be wise: begin!'[16] Then come the lines about the churl and the gliding stream.

Here too the classical and biblical impulses are inseparable. While keeping the skin sleek recalls Horace, the preceding verse is closest to St Paul, warning Timothy that in the last days will come dangerous times, when mankind will be 'lovers of voluptuousness more than of God' ('voluptatium amatores magis quam Dei').[17] Yet in the fifth strophe as in the second the Archpoet also fuses his Horatian thought with one from Job—the despairing, or cynical, thought that fat cats and lean ones meet the same end (21: 23–6):

iste moritur robustus et sanus dives et felix . . .
alius vero moritur in amaritudine animae absque ullis opibus
et tamen simul in pulverem dormient et vermes operient eos

One man dies robust and healthy, rich and happy . . .
another dies in bitterness of soul, without any resources,
yet both will sleep in the dust with worms for covering.

[15] *Gedichte*, 9. 15. Here the form of wording itself decisively indicates a deliberate echo by the Archpoet. There are various other passages in Roman poetry that feature expressions of the *cutem curare* kind: A. Otto, *Die Sprichwörter und sprichwörtlichen Redensarten der Römer* (Leipzig, 1890), 104, compares Horace, *Ep.* 1. 4, 15 and *Sat.* 2. 5, 38, Persius 4. 18 and Juvenal 2. 105, concluding 'Jedenfalls ein vulgärer Ausdruck'. Yet in none of these other classical verses is the context as relevant, or the expression as close, to the Archpoet as in *Ep.* 1. 2. 29. The fact that other medieval writers could adduce the Horatian phrase as a *sententia*—cf. R. Häussler, *Nachträge zu A. Otto* (Hildesheim, 1968), 151; H. Walther, *Proverbia sententiaeque Latinitatis Medii Aevi* (6 vols., Göttingen, 1963–9), no. 11722—in no way precludes its imaginative intertextual use by the Archpoet.

[16] *Ep.* 1. 2. 27–41.

[17] 2 Tim. 3: 4.

The Archpoet inclines to Job's inference rather than Horace's. The words *morior* and *anima* would not in themselves be enough to make a reminiscence likely here; again it is the context, with its contrast between two modes of life, and the nihilistic sense that in death neither will have the advantage, that establishes the bond. Besides, the Archpoet had set a closely similar phrase from Job prominently—almost like an epigraph—at the opening of his poem, transforming the prophet's 'I shall speak in the bitterness of my soul' ('loquar in amaritudine animae meae') into 'In bitterness I shall speak to my own mind' ('in amaritudine loquar mee menti').[18]

My next illustration may seem more surprising than one from Horace: it concerns the much more rarely read Tibullus. (Where the Archpoet will have been able to encounter him is a point I shall come back to later.) The Archpoet's pursuit, as he confesses, is not Horatian wisdom but rather—in order of appearance—sex, gambling, wine, and poetry.[19] When he first mentions sex, the words are:

4 quicquid Venus imperat labor est suavis,
que nunquam in cordibus habitat ignavis.

Whatever Venus commands is a soft hardship—
she never dwells in hearts that are cowardly.

Four strophes later it is the imperial city of Pavia, a city renowned for its hedonism[20] already before Barbarossa made it his Mecca,

8 ubi Venus digito iuvenes venatur,
oculis illaqueat, facie predatur.

where Venus hunts the young with beckoning finger,
ensnares them with her eyes, captures them with her looks.

In Tibullus' *Elegy* I. 2, a section of some ten distichs is devoted to the motif 'You must do and dare, for Venus helps the brave' ('audendum est: fortes adiuvat ipsa Venus'). While this line of Tibullus has not been noted by the Archpoet's editors in relation

[18] *Gedichte*, 10. 1; Iob 10: 1. While the MSS of *Estuans intrinsecus* are fairly evenly divided between *loquar* and *loquor* here, and Bischoff, *Carmina Burana*, i. 3 (Heidelberg, 1970), 6, prefers the second, the biblical echo still seems to me to make the first reading likelier.

[19] Admittedly poetry already receives a passing mention in connection with gambling (10. 10); and the 'tower of Truth' (*turris Alethie*, 10. 9), if we recall the situation in the *Ecloga Theoduli*, must imply, beyond the concept of chastity, that of truthful—as against lying—poetry: Alethia's song, not that of Pseustis.

[20] See esp. the passage from Landulfus of Milan (*c.*1100) adduced by K. Langosch, *Hymnen und Vagantenlieder* (Basle, 1954), 335.

to strophe 4, Watenphul-Krefeld have referred to several lines among those that follow. Even if the notion that Venus never dwells in cowardly hearts exists elsewhere in Roman love-poetry, what is remarkable is its combination, in the same context, with allusions to the gestures and glances by which love can be signalled. In this elegy, Venus not only helps those who are brave (16), but she teaches a woman in love how to give eloquent looks ('nutus . . . loquaces') to her lover, even if her husband is there (21). This leads Tibullus back to the theme of daring (23 f.):

nec docet hoc omnes sed quos nec inertia tardat
  nec vetat obscura surgere nocte timor.

Her teaching's not for everyone but only those with courage
and capacity to rise in the shadow of the night.

Then, a few lines later (33–6), it is Delia, having learnt Venus' lesson, who uses her finger to beckon the poet:

non labor hic laedit, reseret modo Delia postes
  et vocet ad digiti me taciturna sonum . . .
    . . . celari vult sua furta Venus.

No harm in present hardship, if my Delia turns the lock
and calls me silently—with one click of her finger . . .
the thefts of Love are secret—that is Venus' will.[21]

Again these verses have not been mentioned by the Archpoet's editors, and indeed the sense of secrecy in love that underlies the Tibullan moment means that the thought moves differently from the Archpoet's. Yet his elements are Tibullan: the lover is summoned with a finger, or with eloquent looks; and (I would add) it is Tibullus' *non labor hic laedit* that is echoed in *labor est suavis*, and Venus' will that becomes 'whatever Venus commands' ('quicquid Venus imperat'). Before we leave Tibullus, one moment in a later elegy deserves attention too. Tibullus imagines himself addressing Delia's husband, telling him of all the ruses he had used and saying, 'forgive me who am confessing: it was Love that commanded me' ('ignosce fatenti. iussit Amor').[22] While *iussit Amor* is part of the love-poets' *koinê*, I wonder whether *ignosce fatenti* is not echoed in the Archpoet's address to his patron at the close of the Confession:

[21] Here and above, the English is cited from Guy Lee's admirable bilingual edition, *Tibullus*: *Elegies* (Cambridge, 1975). In the remainder of this essay all translations are my own.

[22] The expression *ignosce fatenti* also occurs in Lucan 5. 777, yet there the tone and context are much further from the Archpoet.

24 . . . parce penitenti,
fac misericordiam veniam petenti,
et da penitenciam culpam confitenti.

. . . spare the penitent,
be merciful to him who seeks for pardon,
give a penance to him who confesses his guilt.

Alone, the parallel is so slight that it could be thought mere coincidence; in the light of the verses from the earlier elegy, I am inclined to think it is more.

My next illustration is from one of the best known strophes in the Archpoet's Confession. It is a strophe that was later adapted to open a new composition—a cheerful but relatively unsophisticated drinking-song.[23] Here it might seem as if nothing were left to discover about the Archpoet's fusion of Ovid and the Bible:

12 Meum est propositum in taberna mori,
ut sint vina proxima morientis ori:
tunc cantabunt letius angelorum chori
'sit deus propitius huic potatori'.

It is my resolve to die in the tavern,
so that wines may be near my lips as I die:
then the choirs of angels will sing more joyfully
'May God be merciful to this drinker'.

With the alteration of one syllable—*peccatori* to *potatori*—the cry of the repentant publican in the parable (Luke 18) becomes an imprecation of angels over an unrepentant drinker, a drinker who wants his death to be the consummation of his Dionysiac life in the very way that Ovid wanted his to be the consummation of his love-life (*Amores* 2. 10. 29–38):

Felix quem Veneris certamina mutua perdunt! . . .
cum moriar, medium solvar et inter opus,
atque aliquis nostro lacrimans in funere dicat
'conveniens vitae mors fuit ista tuae'.

Happy the man who dies in the duel of love! . . .
When I die, let me fade in the middle of the act,
so that someone weeping at my funeral may say
'This death was appropriate to your life'.

Yet there is a second sacred moment, not noticed by scholars, that gives a new dimension of meaning to the biblical-Ovidian

[23] Cf. *Carmina Burana*, i. 3 (n. 18 above), 19 f.

transpositions. Why is it not the publican himself but choirs of angels who sing his words, and what is the force of the Archpoet's comparative *letius*, more joyfully? Three chapters earlier in Luke come the parables of the lost sheep and the lost drachma—images of the one sinner who repents. The second ends (15: 10):

Ita dico vobis
gaudium erit coram angelis Dei super uno peccatore penitentiam agente.

So I tell you
there will be joy among God's angels over one sinner who does penance.

If for Luke the angels rejoice over a repentant sinner, for the Archpoet they rejoice even more—*letius*—over an unrepentant one. That is, the poet imagines his angels taking the view of Ovid's mourner: 'conveniens vitae mors fuit ista tuae'. Thus we see the full extent of the mischief and the artistic cunning of the Archpoet's strophe: it is by a supersubtle montage of two moments from Luke and one from Ovid that he undermines the notion of repentance in its respectable Christian sense.

My last illustration again involves a fusion of classical and biblical diction. The data here are not new; the principal ones were supplied in exemplary fashion in Watenphul-Krefeld's edition. None the less I should like to modify their assessment of the situation, and to add one or two details that escaped them.

16 Unicuique proprium dat Natura munus:
ego nunquam potui scribere ieiunus;
me ieiunum vincere posset puer unus;
sitim et ieiunium odi tamquam funus.

17 Unicuique proprium dat Natura donum:
ego versus faciens bibo vinum bonum
et quod habent purius dolia cauponum:
tale vinum generat copiam sermonum . . .

18 Nasonem post calices carmine preibo. *v.l.* calicem

19 Mihi nunquam spiritus prophetie datur *v.l.* poetrie
nisi prius fuerit venter bene satur;
dum in arce cerebri Bachus dominatur,
in me Phebus irruit et miranda fatur.[24]

[24] On the rightful place of the strophes 10. 14–19 (which some scholars have wanted to excise here, because they were originally composed for poem 4), see esp. *Carmina Burana*, i. 3 (n. 18 above), 12. Moreover, it seems to me worth noting the Archpoet's predilection for the 'perfect' numbers 33 and 100: poem 4 has 33 strophes, 5 has 100 lines, 7 has 33 lines, 10 (if left unmutilated) has 100 lines. In the editions, 9 has 34 strophes; yet one of these

Natura gives to each his own gift:
I could never write when fasting;
a child unaided could outdo me when I fast;
thirsting and fasting I hate like death.

Natura gives to each his own endowment:
I, when making verses, drink good wine,
the purest that the hosts' barrels contain:
such wine engenders copious eloquence . . .

After wine-cups I'll outstrip Ovid in song.

Never was I given the spirit of prophecy
unless first my belly was well satisfied;
when Bacchus rules in the brain's citadel,
Apollo streams into me and utters wondrous things.

The closest parallel to the opening of the Archpoet's strophes 16 and 17, as Watenphul-Krefeld recognized, is in an elegy of Propertius (2. 22. 17 f.):

> Unicuique dedit vitium Natura creato:
> mi Fortuna aliquid semper amare dedit.
>
> Natura gave to each created being a vice:
> to me Fortuna gave being always in love.

The initial *Unicuique*, the quasi-personified Natura, and the contrast between what is given to each and what is given to me, to the 'I' (*ego*, *mi*) who in both poems opens the next verse, all make it highly unlikely that this parallel is fortuitous. At the same time, the Archpoet has deftly superimposed Propertius' words on words of St Paul (1 Cor. 7: 7):

> Volo autem omnes homines esse sicut me ipsum
> sed unusquisque proprium habet donum ex Deo
> alius quidem sic alius vero sic.
>
> I should like everyone to be like me,
> but each has his own gift from God,
> one for one thing, another for another.

Here the nominative *unusquisque* is combined with *proprium* and *donum*—words that we find in the Archpoet's verses but not in those of Propertius. The classical-biblical conflation reveals a special irony if we recall Paul's context: he, like Propertius, uses these phrases in the course of an argument defending sexual love.

(st. 11), which is preserved only in a mid-fifteenth-century MS written in Cracow, looks to me like a poorer version (whether first draft or later imitation) of st. 12. If, as I suspect, 9. 11 is superfluous, 9 would likewise have 33 strophes.

Husband and wife must pay each other their debt; each has rights over the other's body, not over their own. Ideally, Paul says, he would like the married to live as chastely as he—yet at once he concedes that each has his own gift, and that that of the married, for sexual love, is not the same as his own. The irony deepens if we recall further, as Hamacher did in his essay and many of the Archpoet's first audience must have done, that Paul's verse 'unusquisque proprium . . .' also introduced chapter 40 of the *Rule* of St Benedict, the chapter 'on how much to drink' (*De mensura potus*).[25] For Benedict, Paul's verse was an *auctoritas* not for sex but for wine: only a small measure, an *emina*—probably a quarter of a litre—per day, and yet, 'if the necessity of place or toil or the heat of summer demand more, let it be at the prior's discretion, always taking care lest satiety or drunkenness creep in'.[26]

Thus the Propertian and Pauline phrases, with their diverse tones and connotations, could lead the Archpoet not only to thoughts of sex but to thoughts of wine. Yet for another classical author, Horace, the thought of wine was wittily bound up with that of poetry. Again Watenphul-Krefeld have noted the most memorable parallels in Horace's *Epistles*, in 1. 5. 19 f.:

> Fecundi calices quem non fecere disertum?
> contracta quem non in paupertate solutum?
>
> Whom have fecund wine-cups not made eloquent?
> Whom have they not made free, even in pinched poverty?

And again, in 1. 19. 2–8:

> nulla placere diu nec vivere carmina possunt,
> quae scribuntur aquae potoribus . . .
> laudibus arguitur vini vinosus Homerus;
> Ennius ipse pater numquam nisi potus ad arma
> prosiluit dicenda.
>
> No poems can please for long, or go on living,
> if they're written by water-drinkers . . .
> From Homer's praise of wine we know he was a tippler;
> even Father Ennius never burst into tales of arms
> unless well-wined.

Yet a sceptic might still wonder whether, since this is a topos of some diffusion, these Horatian parallels are not of a general

[25] Hamacher (n. 6 above), 165.
[26] *Règle de Saint Benoît*, ed. P. Schmitz (Corpus Christianorum, 5th edn., 1987), xl 5.

nature only. Can we be sure that the Archpoet had precisely these passages in mind?

I think there are some small but revealing positive indications which have gone unnoticed. At the opening of the Archpoet's strophe 19 we have the same syntagma as at Horace 1. 19. 7: 'numquam nisi'. I am not suggesting conscious imitation here; none the less, in the use of this syntagma we have a hint of how deeply the Horatian moment had been assimilated by the Archpoet's memory and imagination. So, too, with the word *calices* in Horace 1. 5. 19: alone it would hardly be significant; but in the Archpoet's *conjointure* it is another signal of latent memory. At the close of his strophe 18 comes the verse: 'Nasonem post calices carmine preibo'. Twelve manuscripts of the Confession have *calices*, one *calies*, and six the singular *calicem*.[27] I am inclined to follow Bischoff in accepting the plural *calices*—less because of the weight of manuscripts than because of the poetic weight of Horace.

In the first line of strophe 19 a similar textual problem arises. Five manuscripts have *spiritus prophetie*, which is the phrase in John's Apocalypse, as well as in another of the Archpoet's compositions.[28] But twelve manuscripts have *spiritus poetrie*, and one, *poesie*. I incline slightly towards *spiritus prophetie*, for then, in his fusion of Horace with St John, the Archpoet is also outdoing the pagan poet. If Father Ennius never burst into heroic verse before being well-wined, the new poet is never granted the spirit of prophecy before being sated. In earlier strophes, with witty boastfulness, he had claimed that his was inspired composition, inspired as the verses of well-behaved, sheltered teetotallers could never be. In the Apocalypse, the angel tells John that his writing is the true words of God—'haec verba vera Dei sunt'—and a moment later adds, 'For the witness that Jesus gave is the spirit of prophecy'. Whether the Archpoet has here taken over John's expression direct or has lightly changed it, he is claiming the inspiration of a *vates*, a poet-prophet who is simultaneously Dionysiac and Apolline: it is the god Phoebus Apollo who enters him and speaks in him.

The humour with which the Archpoet characterizes his poetic-prophetic spirit owes something, I suspect, to a jest based on Terence. Where in Terence's *Eunuchus* Chremes cites a saying, 'Without Ceres and Liber, Venus grows cold'—i.e. without bread

[27] *Carmina Burana*, i. 3 (n. 18 above), 18.
[28] Apoc. 19: 10; *Gedichte*, 4. 15.

and wine, love grows cold—in a collection of model letters of about 1220 an Oxford student, writing home for money, has the variant 'sine Cerere et Bacone frigescit Apollo' ('without Ceres and Bacchus, Apollo grows cold'—the writer may also be punning on the medieval Latin *baco* 'bacon', a word of Germanic origin).[29] Unless this variant of the Terentian proverb was invented by the rhetorician—and that seems to me unlikely—this will have been the phrase in the Archpoet's mind as he vaunted his own Apolline inspiration.

In no fewer than five of his poems (1, 2, 4, 7, 10) the Archpoet styles himself *vates*, once even (2. 59) *vates vatum*. It is one of the many traits that link him with the equally brilliant—though more sombre and vindictive—poet of the previous generation, Hugh Primas of Orléans, who, asserting himself, had spoken 'of our good *vates*, our master, our Primas' ('de nostro bono vate, / de magistro, de Primate').[30] Where occasional indications of date in Primas' poetry suggest he was at the height of his powers in the 1130s and 1140s, the various political allusions in the Archpoet's work point to the early sixties of the twelfth century. In all the nine compositions that survive (8 is a brief fragment only), the Archpoet appears as a supremely accomplished artist and virtuoso—we have nothing of his that could be thought of as early work or beginner's work. In the 1160s, allusions within his poems show him to have been in Salerno (6), Burgundy (2), Cologne (5), and most of all in northern Italy (7, 9, 10). What can we conjecture about his formation and his career before those years?

This question, fortunately, is no longer bedevilled by nationalism, by articles with titles such as 'Der Archipoeta war ein Deutscher!'[31] that appeared in the 1930s. We know the Archpoet was born north of the Alps (when writing from Italy, he once calls himself *trasmontanus*),[32] but nothing tells us exactly where. Yet the relations between his poetry and that of Hugh Primas are in certain respects close enough, in my view, to suggest that at some stage, perhaps around 1150, he was a disciple of Primas at Orléans.

[29] *Eun.* 733 (*Sine Cerere et Libero friget Venus*); MS BL Add. 8167, fo. 104r (cit. C. H. Haskins, *Studies in Mediaeval Culture* (Oxford, 1929), 10). Cf. also Otto (n. 15 above), 366; Häussler (n. 15), 120 (the verse is cited by Jerome, *adv. Iovin.* 2. 7); Walther (n. 15), no. 29670. Petrus Cantor, PL 205, 331D, has the variant *Sine Cerere et Baccho friget Venus*; cf. A. Gagnér, *Florilegium Gallicum* (Lund, 1936), 137.

[30] Wilhelm Meyer, *Die Oxforder Gedichte des Primas* (1907; repr. Darmstadt, 1970), 15. 91 f. The citations below are from this edition.

[31] K. Langosch, *Historische Vierteljahrsschrift*, 30 (1936), 493–547.

[32] *Gedichte*, 3. 14.

In the more outward aspects of poetic form the Archpoet was not an innovator. Two of the forms that he uses with consummate skill are ones on which his predecessor, Primas, had already set his stamp. The first was octosyllabics with tirade rhyme. One of Primas' most dazzling compositions tells how he came to fall into disgrace (23. 1–6):

Dives eram et dilectus,
inter pares preelectus;
modo curvat me senectus
et etate sum confectus.
Unde vilis et neglectus
a deiectis sum deiectus . . .

I was rich and cherished,
the chosen leader among my peers;
now old age bends me double
and I am wasted by the years.
So, abased and neglected,
even outcasts cast me out . . .

Primas hammers in his two-syllabled rhyme ten times in all, before passing to a new one. The Archpoet was capable of learning from him at the highest artistic level. He chose the same form to evoke a moment when he too claims to have been in disgrace. Thus in the opening strophe of his *Fama tuba dante sonum* (2. 13–16):

ego caput fero pronum
tanquam frater sim latronum,
reus, inops racionum,
sensus egens et sermonum.

I hang my head
as if I were brother to thieves,
guilty, poor in excuses,
lacking sense and words.

This poem is the Archpoet's homage to Primas' *Dives eram.* Yet I would also see it as the attempt of a disciple to outdo his master: even the fact that he begins with a tirade of sixteen perfect rhymes where Primas had begun with ten may be a small indication of this.

Similarly with the virtuoso use of rhymed hexameters to achieve a comic effect, by devising leonine rhymes that extend over more than one word. Primas had perfected this technique, as in the dialogue between the poet and his cloak, which, being unlined, cannot keep out the cold. At the climax the cloak says to the poet (2. 22 f.):

> Conpatior certe, moveor pietate super te
> et facerem iussum, sed Iacob, non Esaü sum!
>
> Indeed I sympathize, I am moved with pity for you
> and I'd do your bidding—but I'm Jacob, not Esau!

The wit of the outrageous rhymes is enhanced by that of the biblical allusion. The threadbare cloak is the smooth-skinned brother, Jacob, not the hairy Esau; and the cloak, like Jacob, is master of the situation; the poet, like Esau, has been cheated of a blessing—the blessing of warmth.

When the Archpoet takes up Primas' theme of cold and poverty, he outdoes him not only in accumulating pathetic details, but in finding still more grotesque rhymes to undercut the pathos, to make comedy out of what is seemingly heartrending (3. 17–19):

> Asperitas brume necat horriferumque gelu me;
> continuam tussim patior tanquam tisicus sim;
> sencio per pulsum quod non a morte procul sum.
>
> Winter's harshness and the freezing cold are killing me;
> I suffer from a ceaseless cough, like a consumptive;
> I feel my pulse and know I'm not far from death.

In the heady extravagances of leonine rhyme and tirade rhyme, and the themes to which these were applied, the Archpoet is close to Hugh Primas and attempting to surpass him. The relationships I am suggesting do not—emphatically not—mean that I can accept the thesis of Gustavo Vinay, that the Archpoet and Hugh Primas are one and the same.[33] This notion has its source in a garbled note in the chronicle of the Franciscan Salimbene, writing more than a century after the poets. Yet the difficulties about accepting Salimbene's testimony are not merely chronological, as Vinay supposed. They are above all literary. To conflate Primas and the Archpoet is to fail to see, beyond the debts of form and motif that the younger poet shows to the older, the high individuality of outlook, temperament, diction, and tone that distinguishes the two. It is a little as if one were to maintain that Sophocles was Aeschylus, that Shakespeare was Marlowe, or—to cite a verse by Ogden Nash—that 'Thomas à Kempis was Thomas Aquinas'.

[33] G. Vinay, 'Ugo Primate e l'Archipoeta', *Cultura Neolatina*, 9 (1949), 5–40.

In various testimonies Hugh Primas is designated 'of Orléans' (*Aurelianensis*), and much of his life and teaching must be linked with this city.[34] Admittedly in one early chronicle he is also noted as living in Paris, and from his poems we know of his sojourns in several other cities of northern France. Yet I have a special reason for suggesting that it was at Orléans that the link between the Archpoet and Hugh Primas, some twenty years his senior, was forged. It concerns a question of textual transmission. I do not think we can explain away the Archpoet's reminiscences of rare poets such as Tibullus and Propertius as simple coincidence; nor—in default of concrete evidence—should we assume that precisely those verses from the two Roman love-poets which the Archpoet recalls happened to be preserved as 'tags'. But the close study, in recent decades, of florilegia and their implications, by scholars of the calibre of Rouse and Munk Olsen, has thrown much new light on the whole problem of the transmission of the rarer authors. In particular, the intricate but to my mind wholly persuasive arguments of Richard Rouse[35] have shown that Orléans in the mid-twelfth century was *par excellence* the centre where one was most likely to encounter a text of the *Elegies* both of Propertius and Tibullus. So I suggest that it is there, as a disciple of Hugh Primas, and with an exceptional range of classical poetry within reach, that we should picture this exceptional craftsman, the Archpoet, as he was moving towards full artistic maturity. To characterize his relation to the classical poets by way of conclusion, I cannot do better than cite some lines from T. S. Eliot's essay 'Tradition and the Individual Talent'.[36] Tradition

compels a man to write not merely with his own generation in his bones, but with a feeling that the whole of the literature of Europe . . . has a simultaneous existence and composes a simultaneous order . . . No poet, no artist of any art, has his complete meaning alone. His significance, his appreciation is the appreciation of his relation to the dead poets and artists . . . what happens when a new work of art is created is something that happens simultaneously to all the works of art which preceded it. The existing monuments form an ideal order among themselves, which is modified by the introduction of the new (the really new) work of art among them.

[34] Cf. A. G. Rigg, 'Golias and other Pseudonyms', *Studi medievali*, 3rd ser. 18 (1977), 65–109, at 72–81.

[35] R. Rouse, 'Florilegia and Latin Classical Authors in Twelfth- and Thirteenth-Century Orléans', *Viator*, 10 (1979), 131–60.

[36] T. S. Eliot, *Selected Prose* (Harmondsworth, 1953), 21–30, at 23.

5 Walter of Châtillon and the Greeks

A. C. DIONISOTTI

Few medieval poets present us so squarely with the problem of understanding classical influences as Walter of Châtillon. Readers of his *Alexandreis*, from Walter's time down to our own, have always felt that it is more like a classic than anything produced before it, or indeed after, at least until Petrarch's *Africa*. Moreover, the poem was an immediate and lasting success: upwards of 180 MSS survive, not counting excerpts. It may be tempting, especially for a classicist, to relate these two facts: the classical is good; the more classical, therefore, the better; so the *Alexandreis* was justly and deservedly successful, in the best of all possible worlds. Critics less sympathetic to the classical underline the scholastic success of the poem: schoolmasters liked it, because it exemplified classical diction and metre. But, on both counts, the very different fate of Petrarch's *Africa* shows that there are holes in such reasoning. Schoolmasters were not short of real classical poems, if diction and metre were all that mattered. And in any case the success of the *Alexandreis* was not confined to schools: it was quoted, quarried and imitated in all kinds of literature, both Latin and vernacular. Somehow, we must suppose, the argument of the *Alexandreis* made sense to contemporaries. What sort of sense, how and why, and what role its classicism played in this, are the questions I would like to consider.[1]

Walter himself, in a rare concession to the traditions of Christian epic, wrote a prose preface for his poem. This, however, does little

[1] Quotations are from Galteri de Castellione *Alexandreis*, ed. M. L. Colker (Padua, 1978). The best study of the poem remains H. Christensen, *Das Alexanderlied Walters von Châtillon* (Halle, 1905); see also the English translation by R. Telfryn Pritchard (Toronto, 1986) with useful bibliography, and H. Harich, *Alexander Epicus: Studien zur Alexandreis Walters von Châtillon* (Graz, 1987). Study of the poem will be much facilitated by H. E. Stiene and J. Grub, *Verskonkordanz zur Alexandreis des Walter von Châtillon* (Hildesheim, 1985), which only became available to me after this paper was written.

to explain the nature and purposes of the work. It concentrates, conventionally enough, on pre-empting the criticism of envious tongues: such critics, Walter pleads, should bear in mind the short time spent on composition, some five years, and the grandiose subject-matter, *altitudinem materiae*, not attempted by any ancient poet. This last point, though it is given more specific meaning in the body of the poem, in itself only echoes the claims to originality made by ancient poets; it does not, any more than for ancient poets, adequately explain the choice of theme: the fact that a theme has not been treated before does not explain why there is any need to treat it now. But the last sentence of the preface abruptly introduces a surprising twist which may be significant (Prol. 39–42):

Sed hec hactenus. nunc autem quod instat agamus; et ut facilius que quesierit quis possit invenire, totum opus per capitula distinguamus.

So much for that, now let's get down to business; and so that the reader may more easily find what he's looking for, let's divide the whole work into chapter-headings.

What Walter means is the short verse summaries that precede each book, like the verse *argumenta* accompanying books of the *Aeneid*. But to call such summaries *capitula*, and explain them as designed for ease of reference, is to relate them to something quite different, namely to serious informative prose; suffice it to say that the archetype for this remark is the preface to Pliny's *Natural History*. Walter himself uses it again in the preface to his prose theological dialogue against the Jews. The work we are about to read, he seems to hint here, is no merely aesthetic or literary entertainment.

This is the only external clue we have. Early commentators, and some modern ones, observe that the poem is dedicated to William of the White Hands, Archbishop of Rheims from 1176, and convert that into the poem's *raison d'être*: the purpose of the poem is to please the Archbishop, or to win his favour. This is good Servian tradition, but it does not explain why the poem should be about Alexander the Great, why it is the kind of poem it is, or why it was so widely successful. In fact it explains nothing at all.

If we try to place the poem in the literary landscape of the time, we unearth a lot of questions but no ready answers. The history of Latin epic in the twelfth century still seems to be a subject in search of an author. In Manitius' monumental work there is no such concept; the *Alexandreis* is discussed in the chapter on 'Persönliche Dichtung', wafted there on the coat-tails of Walter's accentual lyric verse, which also rates twice as much space. It is indeed arguable, Romantic presuppositions apart, that Walter had a more assured and refined command of the more modern medium: on the level of style the *Alexandreis* can let one down with a bump; hyperbole, conceit, and word-play are sometimes misjudged, even by the generous standards of Lucan or Claudian. But epic is not only style. It requires scale, plot, character, speech and action, tradition and invention, all welded to create a large and imaginative and not merely individual view of the human condition. What immediate precedents had Walter for such an undertaking, if any? The poems that most readily spring to mind, the *Ligurinus*, Joseph of Exeter's *Ylias*, and (for all the differences) the *Anticlaudianus* and the *Architrenius*, were all composed after the *Alexandreis*; soon after, within a generation: so did Walter's poem set a new standard, or was it itself just an early example of a wider trend? Large-scale narrative Latin poems of earlier date are not lacking, but it remains to be seen how far they constitute a developing tradition: one feature that most of them share is minimal diffusion.

Further questions arise if we peer over the linguistic fence to consider developments in the vernacular. One of the few bits of biographical information about Walter that cannot be a deduction from his poems is the statement that he was taught by Stephen of Beauvais, whom it is plausible to identify with the Stephen described by Helinand as 'de clericis Henrici comitis . . . exercitatissimus in omni genere facetiarum utriusque linguae, latinae et gallicae'. Count Henry of Champagne was the elder brother of William of the White Hands;[2] but in any case it is scarcely credible that Walter was unaware of the *romans antiques* composed in rather sudden profusion since the 1150s, including some about Alexander the Great. In what measure is the *Alexandreis* a response to these, and with what implications?

Such questions may offer us a framework, even if we cannot answer them. But the first essential is to look at the poem itself, and see what it can tell us. It is a poem of some five and half thousand lines, in ten books, giving a selective account of Alexander's career from his accession to his death. Of course it is a

[2] J. F. Benton, 'The Court of Champagne as a Literary Center', in *Speculum*, 36 (1961), 551–91. One of Walter's lyric poems (below, n. 17, poem 4) may have been dedicated to Count Henry.

career of conquest. After rapid suppression of mutiny in Greece, Alexander crosses to Asia Minor, inflicts an initial defeat on the Persians, briefly sorts out Palestine and Egypt, then concentrates on the major business of wiping out the Persian Empire; in two great battles, Issus and Arbela, he defeats the Persian King Darius, who escapes only to be murdered by his own nobles. Alexander pursues the murderers eastwards, quickly conquers the Scythians (noble savages who had tried to suggest he might live and let live), then confronts the army of King Porus of India, and defeats that. At the beginning of Book 10, as Alexander sets out to impose his power on the ultimate Ocean that bounds the earth, personified Natura, aghast that this human creature threatens to overstep the limits of creation, goes to enlist the powers of hell against him. Personified Betrayal (Proditio) duly prepares the cup of poison that will kill him in Babylon, though not before he has received volunteered submission from the entire western world, from Spain to the Rhine.

Alexander, then, is the unrivalled central figure. But what kind of hero is he, and what is the purpose of telling his story?

The criteria by which Walter selected and adapted his sources may give us some bearings. By the late twelfth century, there was no shortage of Latin material about Alexander. Besides the brief accounts in Orosius and Josephus, digested in Peter Comestor's *Historia Scolastica*, there were the historians Justin and Curtius Rufus, and several texts deriving from the Greek Alexander romance, both handed down from late antique versions and more recently translated from Eastern sources.[3] Moreover, as Cary beautifully demonstrated, Alexander's character and the significance of his career had been much discussed by theologians in the wake of patristic judgements. There is evidence in the poem that Walter was familiar with most of this material, but his selection from it is single-minded and highly abstemious. A few examples will suffice.

From an early period, all manner of stories clustered round the birth and youth of Alexander—notably that he was not really the son of Philip, but of an Egyptian astrologer-king Nectanebus, who seduced Philip's wife Olympias by visions and transformation into a serpent. This not only occupies much space in various versions of the romance, it is also referred to in Justin's history, as

[3] Surveyed and discussed in G. Cary, *The Medieval Alexander* (Cambridge, 1956) and D. J. A. Ross, *Alexander Historiatus* (Warburg Institute Surveys, 1: London, 1963; 2nd edn., Frankfurt a.M., 1988).

a story that the adult Alexander is at pains to quash by consulting the oracle of Jupiter Ammon. It gets no such showing in the *Alexandreis*. Early in the first book, the twelve-year-old Alexander dismisses it in all of four words (1. 46–7 'semperne putabor / Nectanebi proles?'), and thereafter it is only used for the odd bit of malicious innuendo on the part of Alexander's adversaries.[4]

Likewise, Walter's Alexander meets no basilisks, talking trees, or fire-breathing monsters on his eastern journey. When he gets to India he joyfully exclaims (9. 203–5):

> 'Inveni tandem dignumque stupore meoque
> par animo discrimen,' ait 'res ecce gerenda est
> cum monstris michi cumque viris illustribus una.'

—at last he has found a truly amazing contest worthy of himself, a battle with monsters and famous warriors all at once; but the monsters are in fact elephants, monstrous in their size, but otherwise portrayed as fully natural, and naturally vulnerable.

Walter evidently also knew the story of Alexander's journey to paradise, but he only alludes to it, and very cleverly, in the explicitly mythological tableau in Book 10, where Natura thus clinches her plea to Lucifer (10. 102–4):

> que tua laus, coluber, vel que tua gloria primum
> eiecisse hominem, si tam venerabilis hortus
> cedat Alexandro?

Where's your pride and glory, snake, your expulsion of Adam, if that so venerable garden falls to Alexander?

Thus Walter alludes to the fabulous legends about Alexander, but systematically and rigorously excludes them from his narrative. Was it because he thought such stories were too folk-loreish, not suitable for epic style? But there are plenty of epic models for handling portents, visions, magic, and the like. Was it because he did not believe these stories to be true, and was concerned to give a bona fide historical account? This at first sight seems more plausible, fitting in with the fact that he chose Curtius, the fullest and most soberly historical account available to him, as his primary source, followed very closely indeed for large stretches of his narrative. But I do not think historicity is the whole answer either; for Walter both invents much quite un-historical matter, and omits or dismisses matter that was both acceptably historical and potentially heroic.

[4] 2. 333, 3. 167, cf. Christensen, 141 n. 1.

A revealing case is the very opening of the poem. The narrative proper begins, as I mentioned, with Alexander's accession to the throne, imagined by Walter as a peaceful coronation at Corinth. But before that, immediately following the dedication of the poem, there is a curious scene, a sort of play before the play, presenting the youth of Alexander. We hear nothing of the taming of Bucephalus, of the victory at the Olympic games, or of Philip's murder and Alexander's avenging of it. Instead, we are given two speeches, not prompted by any particular circumstance: a monologue by the twelve-year-old Alexander, exasperated by his own youth and impatient for action and victory; and then a long lecture to him by Aristotle, on the arts of power and rule. Whether for epic or history, the episode is strangely timeless, symbolic, adrift from narrative.[5] If we opened the book expecting the adventures of Alexander, or even a historical biography of him, this is clear warning that the poem will be neither. Kingly power is its theme; that is what governs the poet's choice and use of both history and imagination.

Turning to look at the poem as a whole, we see that Walter has taken pains to set his story into a very large historical and geographical perspective. Among the episodes entirely invented by him are two lengthy descriptions of tombs, those of Darius' wife and of Darius himself, both sculpted by a Jewish artist in Alexander's service called Apelles. The first tomb depicts the history of the world from the Creation down to Alexander, based, of course, on the Old Testament. The second has a canopy representing a map of the world with all its lands and peoples; it also has an epitaph for Darius and a date calculated from the Creation, which reassert the historical dimension.[6] Elaborate descriptions of works of art are a traditional feature of epic, and critics have rather dismissed Walter's tombs as artificial and anachronistic bits of epic machinery. Their purpose, however, is quite clear: Alexander is not just one epic hero among many, he is the culmination of human history to date, in its central development as established by the Old Testament, not in some

[5] Walter may have taken a hint from Lambert Le Tort's Alexander poem, which also began with a protreptic speech by Aristotle, albeit more closely keyed into the narrative and focused on the dangers of promoting low-born servants (E. C. Armstrong et al. ed., *The Medieval French Roman d'Alexandre*, ii (Elliott Monographs, 37: Princeton, 1937), 143–5). Walter's scene as a whole owes more to Claudian (besides the references given by Colker cf. *Paneg. IV cons. Honorii* 358–78 with *Alexandreis* 1. 33–47, 198–9).

[6] 4. 176–274, 7. 379–430; cf. also 1. 396–426, 2. 494–539.

side-show like the Trojan War. And the whole known world is his stage, extending from Arthur's Brittany to India.

But if the range of time and space is about as large as it could be, the range of human relationships depicted in the poem is extraordinarily narrow. Beyond the glimpse of the boy Alexander as pupil to Aristotle, we only ever see him as a commander of men. He inspires, and to a degree reciprocates, loyalty in his troops, and his dealings with his officers successively illustrate devotion and betrayal, on both sides. Alexander behaves with impeccable but distant courtesy towards Darius' mother, wife, and children, who are his prisoners of war. Darius is so touched by this that he offers Alexander peace, a share of his kingdom, and a daughter in marriage. Alexander rejects this: he is courteous to Darius' family, he coldly explains, because he has not come to fight against women and children (4. 144–51). It is as if anyone who cannot fight, either with him or against him, scarcely exists.

Walter's Alexander is in fact much lonelier than the supposedly lonely hero Aeneas. He has no father-figure like Anchises or Evander. He not only has no wife or son, he never displays the slightest interest in acquiring any. Roxane is never mentioned. And, in this age of burgeoning romance, this poem has no romantic episode. Ancient tradition and modern versions offered the exotic Queen Candace, who lured Alexander to her castle and almost held him in thrall; Walter ignores the story. All we get is the Queen of the Amazons, who comes to meet Alexander, in full armour and with 200 of her cavalry, as he journeys east. Though initially disappointed with Alexander's physique, she none the less proceeds to a business-like proposal, monarch to monarch, for a bit of genetic planning (8. 36–48):

> ergo rogata semel ad quid regina veniret
> anne aliquid vellet a principe poscere magnum,
> se venisse refert ut pleno ventre regressa
> communem pariat cum tanto principe prolem,
> dignam se reputans de qua rex gignere regni
> debeat heredes. fuerit si femina partu
> prodita, maternis potietur filia regnis.
> si mas extiterit, patri reddetur alendus.
> querit Alexander sub eone vacare Talestris
> militiae velit. illa suum custode carere
> causatur regnum. tandem pro munere noctem
> ter deciesque tulit, et quod querebat adepta
> ad solium regni patriasque revertitur urbes.

She would like to have a child by him, preferably a girl: if it turns out a boy, he can keep it. Alexander reacts by suggesting she might like to join his army—presumably it would save time. When she objects that she cannot abandon her kingdom for so long, Alexander finally ('tandem'), agrees to spend some nights with her, thirteen to be precise, and the queen, having prosily got what she wanted ('quod querebat adepta') returns to her kingdom. Here Walter is simply versifying the dry account in Curtius. That is no answer, however, for he is quite capable of expanding and transforming Curtius when he wants to. Thus he turns a brief notice of an exploit by two officers into a complete re-run of the Nisus and Euryalus episode of *Aeneid* 9.[7] Clearly he wants no Dido and Aeneas, not even any Cornelia and Pompey, or Caesar and Cleopatra.

In the Alexander legend, Alexander's mother Olympias is a prominent figure to whom he is warmly devoted. In Walter's poem she is only rarely and dimly evoked, and one passage starkly redefines the relationship. In the thick of the battle of Arbela the twin gods of war, Mars and Bellona, observe that while Alexander is hell-bent on chasing Darius, his own best regiment is being badly hammered by one of Darius' generals. Bellona, in suitably horrendous Gorgonic attire, descends on Alexander to tell him that his chase is in vain, a different fate has been decreed for Darius, he should return to help his men. At this Alexander (5. 241–50):

> excutitur saltu Macedo profugamque secutus
> voce deam,'quocumque venis, dea, cardine, vanum
> spernimus omen,' ait 'non me divellet ab armis
> et curru Darii licet impiger ales ab alto
> missus Atlantiades verax michi nuncius ipsas
> afferat a Persis raptas cum matre sorores.
> ex Dario pendet nostri spes unica voti,
> quem si perdidero, parvi michi cetera, parvi
> perdita momenti. solum si vicero, solus
> perdita restituet . . .'

shakes himself with a jump and calls after the fleeing goddess 'Whatever corner of the world you come from, goddess, I scorn your empty omen: I'll not be dragged away from the arms and chariot of Darius, not even if the true messenger Mercury swoops down despatched from on high to tell me that my own mother and sisters have been raped by the Persians.

[7] 9. 77–147.

On Darius depends my one hope and prayer; if I lose him, it little matters what else I lose, if I defeat only him, he alone will make up for any losses . . .'

and more in the same vein. The episode is of course entirely invented by Walter—most of his battle descriptions are freely invented in the epic manner. But we should note that Alexander here is in fact made to break all the rules for how to behave, in epic, when visited by a deity: normally your limbs freeze, your hair stands on end, and your voice sticks in your throat; and you do not proceed to tell the deity to get lost. Walter even seems to underline the abnormality, by the reference to Mercury flying down from on high, complete with verbal echoes of his most famous such flight, when he went to tell Aeneas to stop dallying in Carthage and think of his destiny, at least for the sake of his son. For Alexander, the pursuit of victory, as defined by himself, is all: to hell with what fate may have decreed, to hell with his mother and sisters.[8]

What then of Alexander's character? Are we supposed to admire, or abhor, this drastically single-minded pursuit of pre-eminence and victory, which is his dominant characteristic throughout the poem? It is of course a commonplace, in both biblical and pagan moralizing tradition, to condemn such vaulting ambition, such vanity of vanities. But we should hesitate before concluding that the poet devoted himself to writing ten books of highly original epic merely to reassert this commonplace. And in fact such a view fails to account for one of the poem's most striking novelties.

The entire Alexander tradition had always stressed the dramatic degeneration of Alexander's character: the virtues of courage and leadership that brought him success were corrupted by that very success; oriental wealth and luxury bred self-indulgence, intemperate and tyrannical behaviour; Alexander had conquered the Persians, ran a much quoted epigram, only himself to be conquered by Persian vices. Curtius has several long disquisitions on this theme, and offers plenty of graphic evidence for it in Alexander's actions.

Walter does versify one such disquisition in Book 3, commenting on Alexander's generous treatment of Darius' family: if only he had continued like that, no one would have had cause to denigrate him; but Persian wealth and success corrupted him (3. 250–2):

[8] Cf. also 2. 381–7, 3. 280–1, 4. 361–6.

qui pius ergo prius erat hostibus, hostis amicis
impius in caedes et bella domestica demum
conversus, ratus illicitum nichil esse tyranno.[9]

So whereas before he was saintly to his enemies, he became an evil enemy to his friends, and finally turned to murder and war against his own people, and thought like a tyrant that nothing was unlawful for him.

But in the remaining seven and a half books of the poem, Walter systematically omits all evidence for this transformation of Alexander's character. At the beginning of Book 6, Babylon is invited to admire her new king (6. 8–15):

rex erit ille tuus a quo se posceret omnis
rege regi tellus, si perduraret in illa
indole virtutum qua ceperat ire potestas.
aspice quam blandis victos moderetur habenis,
aspice quam clemens inter tot prospera victor,
aspice quam mitis dictet ius gentibus, ut quos
hostes in bellis habuit cognoscat in urbe
cives, et bello quos vicit vincat amore.

Your king will be the sort of king by whom every land would wish to be ruled, if his power remained set in the virtuous nature in which it first grew. See with what gentle reins he governs the vanquished, see how merciful is the victor amidst all his success, see how in the city he recognizes as citizens those who were his enemies in war, and conquers with love those whom he conquered by war.

This is not in Curtius. Babylon is of course a pit of corruption, and the poet reverts to Curtius for a general statement that Babylonian luxury broke down Alexander's innate and acquired virtues. But in the following narrative it is not Alexander who is corrupted, but his troops; or rather, they would have been, except that Alexander promptly introduced a programme of moral re-armament, with a reform of military organization which Walter goes so far as to argue was a greater achievement than the whole of Roman Law. In Curtius' account the capture of Persepolis which follows is celebrated by a drunken orgy that gets out of hand and leads to the wholesale destruction and burning down of the famous city. In Walter's version the capture and the destruction are rolled into one, there is no mention of any orgy, and the city's fate is emotionally justified by only then recounting, as a flashback, Alexander's piteous meeting with a convoy of Greek prisoners, brutally mutilated by the Persians.

[9] Cf. Curtius 3. 12. 18–21 and Justin 11. 11.

Famous among the crimes of Alexander were the murders of his friends, especially Clitus and the philosopher Callisthenes. Curtius describes them at length, underlining the corrupted Alexander's jealousy, intemperance, arrogance, and cruelty. Walter just registers their deaths in a couple of lines (9. 3–8):

> . . . quam [sc. Indiam] dum petit ille deorum
> aemulus in terris, Clytus, Ermolaus et eius
> doctor, Aristotili preter quem nemo secundus,
> extremum clausere diem, documenta futuris
> certa relinquentes: etenim testatur eorum
> finis amicitias regum non esse perennes

and draws a quite different moral: their end bears witness for future generations that the friendships of kings do not last. Not Alexander's character, then, but a general truth about kings.

As for tyranny, right at the end of the poem, when Alexander receives the homage of all Western peoples, Walter gives him an invented speech. These people have nothing to fear from his rule, says Alexander (10. 289–98):

> cui si se Darius posito diademate supplex
> commisisset, eo regnorum in parte recepto,
> sensisset nichil esse iugo mansuetius isto.
> Porus in exemplo est qua mansuetudine victis
> presideam victor, nedum parentibus ultro.
> quosque iugum nostrum vis nulla subire coegit
> subiectos michi mortales ita vivere salva
> libertate volo ut iam non sit servitus, immo
> libertas, servire michi. distinctio nulla
> libertatis erit inter quos nemo rebellis.

Had Darius surrendered, he would have had a share of the kingdom and found no yoke more gentle than Alexander's; Porus is an example of how gently he treats the conquered, never mind those who submit voluntarily. Such people will be subject to him in such a way as to retain their full freedom, it will be no servitude, indeed to serve him will itself be freedom; where there is no rebellion, all will be equally free. There is a disquieting sense of Antichrist here, of a demonic man usurping the role and language of the Christian god. But it is a far cry from the conventional picture of the corrupted tyrant. And Alexander goes on to make a speech of quite exemplary generosity to his soldiers.

In all this I do not think the poet's purpose is to glorify or whitewash Alexander: what he does is rather to isolate his single driving passion, his thirst for unchallenged power, stripping it bare of the conventionally concomitant vices that would make moral judgement easy. Wealth, luxury, love, family, friends, one's own country, all these matter, for good or ill, to other men in the poem, not to Alexander. He is savagely ruthless against any disloyalty or opposition, real or supposed, but otherwise generous, just, courageous to a fault, and brilliant in the art of war. The upshot is a poem not so much about Alexander the Great but, as the preamble implied, about political and military power; power that requires a combination of passion and skill, and extremes of both self-obsession and self-denial.

Is such power a good or a bad thing? For us in the twentieth century, with our democratic traditions and autocratic realities, the notion of power like that of Walter's Alexander is rather automatically abhorrent; and the warrior heroism of epic tradition is now so alien to serious art that we may be tempted to interpret it away as ironic anti-heroism, or schematize it into a tussle between good and bad; either way, to seek a moral excuse for it. But Walter was no democrat, and the very stuff of his verse shows that he was steeped in the epic tradition. I think that if we respect his text, its tone, proportions and emphases, what is said and what is not, we can no more draw a simple moral from his poem than we can from the *Iliad* or the *Aeneid*; in other words it is truly an epic, not a didactic, poem: its aim is to explore a vision at the limits of human greatness and frailty, not to teach an orthodoxy.

This seems to me inherently true of the poem as a whole, but one passage makes the point as explicit as it could well be in epic poetry. Alexander's goddess is Victory, and Victory holds court in a palace (4. 408–32):

ad limina prima susurrat
introitumque tenet curarum sedula mater
Ambitio pernox. solio sedet intus eburno
diva, triumphales lauro mordente capillos,
munifica munita manu, cinguntque sorores
eius utrunque latus et regia tecta coronant
perpetue comites: lirico modulamine carmen
inmortale canens et in evum Gloria vivax,
Maiestasque premens rugoso secula fastu,
concilians que sibi facilem Reverentia plebem,
et dea que leges armat, que iura tuetur,
Iusticia, in neutram declinans munere partem.
assidet hiis stabilitque deae Clementia regnum,

sola docens miseris misereri et parcere victis.
has inter locuples sed barbara moribus astat
fomentum vicii genitrixque Pecunia luxus.
pacifico reliquis prelibans oscula vultu
inmemor est odii finis Concordia belli
et Pax agricola et cum pleno Copia cornu.
Applausus a fronte sedent, qui seria ludis
miscentes vario divam oblectamine mulcent,
et Favor ambiguus et bleso subdolus ore
Risus adulator, commentaque Ludicra divae
singula policronos aptant, et musica circum
instrumenta sonant numeros aptante camena.

At the first threshold sleepless Ambition, anxious mother of cares, whispering guards the entry. Within sits the goddess on an ivory throne, the laurel of triumph gripping her hair, her hand generous with gifts. Her sisters and constant companions surround her and adorn her royal palace: everliving Glory sings an immortal song of lyric melody, Majesty overawes the world with frowning pride, Reverence wins the people to her, and Justice arms the laws and defends rights, unswerving in her task. Sitting by them is Clemency, securing the goddess's kingdom, who alone can teach to take pity on the wretched and spare the defeated. Among them stands Money, rich and vulgar, the nurse of vice and mother of dissipation. Concord with her visage of peace snatches kisses from the rest, forgetting hatred and ending war, together with Peace the farmer and Plenty with her full horn. Before the goddess sits the crowd of Applause, soothing her with varied pleasure, serious and playful, fickle Favour, flattery's Smile and deceitful Shows, all ready to cry 'Vivat' for the goddess, while musical instruments sound to the rhythms of poetry.

The tableau has a long ancestry in classical and classicizing poetry, the personifications suggest moral discourse in epic dress. But this is no psychomachy of orderly virtues marshalled against opposing vices: Walter mythologizes a real world, in which victorious power depends upon both virtue and vice, truth and illusion. The style invokes the moralist, only to defy him in the substance.

Why should Walter have been so interested in such power? It is an immediately striking fact that this poem about military might and world rule is repeatedly and explicitly anti-Roman. This begins in the very first lines (I. 1–11):

Gesta ducis Macedum totum digesta per orbem,
quam large dispersit opes, quo milite Porum
vicerit et Darium, quo principe Grecia victrix
risit et a Persis rediere tributa Corintho
Musa refer. qui si senio non fractus inermi

pollice fatorum nostros vixisset in annos
Cesareos numquam loqueretur fama tryumphos,
totaque Romuleae squaleret gloria gentis:
preradiaret enim meriti fulgore caminus
igniculos, solisque sui palleret in ortu
Lucifer, et tardi languerent Plaustra Boete.

Alexander is the prince under whom 'Grecia victrix risit', victorious Greece smiled. Had he lived on, fame would have had nothing to say of the triumphs of the Caesars, the whole glory of the Roman race would be tarnished, for the furnace with the blaze of its merit would outshine the little flickers, and at the rising of their sun Lucifer would pale and the wagon of plodding Bootes would wane.

Now *Graecia victrix* was a concept that the Romans never really accepted this side of the Trojan War. *Graecia capta* might indeed conquer her rough conqueror with the fine arts, statues of breathing bronze, and the like. But precisely on condition of being *capta*: the arts of conquest and rule were Roman. Alexander made the Romans nervous, but they coped by treating him as a flash in the pan, a monstrous individual, in any case a Macedonian, not a representative Greek. Many writers inveigh against his vices and folly; Virgil more tellingly, in the whole vast canvas of the *Aeneid* going from the Trojan War to Augustus, never once mentions him. But Walter is not merely protesting the importance of his subject against a literary tradition: time and again he uses Alexander to rob Rome of all it stands for. The Eusebian doctrine, that the Roman Empire had been ordained by God to further and maintain Christianity, was the common ground underlying even the most divergent theories of papal versus imperial power; it is excluded from Walter's vision. Instead, half-way through the poem, the peroration of Book 5 delineates a revealing alternative. Alexander's triumph after Arbela, says Walter, was greater than any ever seen by Rome, whether that of Augustus after Actium or Julius Caesar's after Pharsalus. And rightly so, since Alexander's success against such heavy odds and on such a scale makes the heroes sung by Lucan and Claudian look like mere *plebs*. The poet then turns to the future (5. 510–20):

si gemitu commota pio votisque suorum
flebilibus divina daret clementia talem
Francorum regem, toto radiaret in orbe
haut mora vera fides, et nostris fracta sub armis
Parthia baptismo renovari posceret ultro,

queque diu iacuit effusis menibus alta
ad nomen Christi Kartago resurgeret, et quas
sub Karolo meruit Hyspania solvere poenas
exigerent vexilla crucis, gens omnis et omnis
lingua Ihesum caneret et non invita subiret
sacrum sub sacro Remorum presule fontem.

If divine clemency, moved by the tearful prayers of her people, granted France a king like Alexander, then soon the true faith would shine over the whole world. Parthia, broken by French arms, would itself beg for baptism, Carthage long in ruins would rise up again in the Christian faith, and the standards of the cross would exact punishment from Spain for its defeat of Charlemagne. Every land and every people would sing of Christ and willingly seek baptism from the Archbishop of Rheims.

The text warrants a few glosses. Parthia, Carthage, and Spain represent points of the compass and also, of course, Islam. France's Alexander would be no ordinary Crusader setting out to rescue the Holy Places; his aim would be no less than a new world empire to be guardian of the Christian faith, in defiance of the established scheme of universal history which allotted that role to Rome. Moreover, 'Parthia' and 'Carthage' are no idle poetic archaisms. The new French empire would undo Roman history, ancient and modern: it would vanquish Rome's one unvanquished enemy, and resuscitate the enemy that Rome had destroyed, as well as avenge the defeat of Rome's second founder, Charlemagne.

One might usefully compare the Archpoet's one public poem, a hymn to Barbarossa after the destruction of Milan in 1162.[10] It begins, 'Salve mundi domine, Caesar noster, ave!'; the emperor is 'princeps terrae principum', he will restore Roman sway:

qui potenter sustinens sarcinam mundanam
relevat in pristinum statum rem Romanam.

He is both a new Charlemagne in war, 'repraesentat Karolum dextera victrici', and a new Augustus in peace, the peace which heralded the Incarnation and was the cornerstone of the Eusebian doctrine:

iterum describitur orbis ab Augusto,
redditur res publica statui vetusto,
pax terras ingreditur habitu venusto
et iam non opprimitur iustus ab iniusto.

[10] *Die Gedichte des Archipoeta*, ed. H. Watenphul and H. Krefeld (Heidelberg, 1958), 69–73, 127–38 (poem 9).

It is then only natural that Greece, ancient and modern, should be seen as an antithesis. The Greek emperor is admonished to take due warning from the defeat of Milan:

> interim precipio tibi, Constantine,
> iam depone dexteram, tue cessent mine . . .
> tantus erat populus atque locus ille,
> si venisset Grecia tota cum Achille
> in qua tot sunt menia, tot potentes ville,
> non eam subicere possent annis mille.[11]

Walter was no monk buried in some introspective cloister. His patron and perhaps employer was a scion of the house of Champagne, nephew of the ousted King Stephen of England and brother of the Queen of France. Walter travelled to Rome and Bologna, through an Italy scarred by the ravages of Barbarossa. His other poems reveal a keen eye for the events of his day, as do many passages of the *Alexandreis*. But when did he write it? It is generally said that the poem was composed between 1178 and 1182, and published shortly after. For various reasons I think we are much more probably in the early seventies.[12] The current king of France is then the ageing Louis VII, his only son is a sickly boy of between five and ten, who may or may not survive. Louis VII was famous for his great piety, but for little else. His crusade in 1147 had been an ignominious failure; he had lost both his wife and much of France to Henry Plantagenet, who repeatedly outmanoeuvred him. On the other side he was dwarfed by the ruthless conqueror all too like Alexander, the heir of the Caesars, Frederick Barbarossa. A Frenchman might well ponder the nature of political and military power.

The choice of Alexander as a theme gave scope for a poem simultaneously classical, secular, and anti-Roman. But the theme did not of itself necessitate such treatment. When Walter wrote, large-scale poems on ancient themes were all the rage, but written in the vernacular, utterly unclassical in manner and much of their matter, and especially associated with the Plantagenet court. One way to consider Walter's classicism would be as a deliberate challenge and antithesis to this vogue, an assertion of cultural superiority on a comparable scale and ground. Certainly, if one compares the *Alexandreis* with Stephen of Rouen's *Draco*

[11] Cf. also stanza 31 'volat fama Cesaris velut velox equus, / hac audita trepidat imperator Grecus; / iam quid agat nescius, iam timore cecus, / timet nomen Cesaris ut leonem pecus'. And contrast *Alexandreis* 10. 182–4.

[12] See below, Appendix.

Normannicus, completed in honour of Henry II in 1169, one can see why the latter poem was left to lie discreetly uncopied at Bec: some 4500 elephantine Latin elegiacs, reeking of the schoolroom, yet innocent of any classical shape or style, it could be no epic advertisement for the patron of John of Salisbury and Peter of Blois. By the time Joseph of Exeter set about his *Ylias* (1188–90), it was an exercise in catching up: Joseph is clearly inspired and influenced by the *Alexandreis*.

In a different perspective, the secular and anti-Roman character of Walter's poem might also be seen as a literary concomitant to the increasingly secular, and increasingly Greek, philosophy burgeoning in the schools of Paris. Walter knew very little about the Greece whose world empire he conjured up to challenge that of Rome and its heirs. But he knew, of course, that Athens is the immortal spring of science and philosophy. When Aristotle appears at the beginning of the poem (1. 59–62), wan and desiccated with study, he has just put the last full stop on his *Sophistici Elenchi*:

> forte macer pallens incompto crine magister
> (nec facies studio male respondebat) apertis
> exierat thalamis ubi nuper corpore toto
> perfecto logyces pugiles armarat eleñcos.

New translations of the logical works were being sought out by inquiring Westerners at this very time; still it is quite something that a text-book of advanced logic should figure in an epic poem, complete with title.[13] The speech that Walter gives Aristotle on the arts of rule is freely composed, and certainly it knows nothing of the *Ethics* and *Politics*. But, like the figure of Alexander himself, it breathes a remarkably secular spirit, irreducibly of this world, the spirit that makes Aristotle so much more of a challenge to Christian thought than Plato. It was not till the next generation that men began to take up the challenge of the *Nicomachean Ethics* and size up the vast reaches of Greek philosophy that Rome, pagan and Christian, had failed to transmit.[14] By then, too, France was seeing an Alexander in Philip Augustus. These were changes in the habits of thought that required much intellectual openness and energy. The *Alexandreis* did not go uncontested;

[13] Cf. also the scientific language at 1. 59–71, 415–17, 2. 166–70.

[14] The more Aristotle was discovered, the more difficult it was to accommodate him in Western tradition and to explain his poor showing in Latin texts, especially in Augustine. Cf. the essentially hostile account in John of Salisbury's *Entheticus* 823–936 and in Alan of Lille's *Anticlaudianus*; a century later Roger Bacon could bluntly conclude that Augustine had just not read enough Aristotle.

indeed the *Anticlaudianus* is one kind of answer to it. But it remained a classic of that new readiness to break out of the Western mould, to question the boundary and to declare, like Alexander, that one's own world is scarcely enough.

Appendix
On the Dating of the *Alexandreis*

The starting-point for dating the *Alexandreis* has always been its dedication to William of the White Hands, who became archbishop of Rheims in 1176. That the dedication was no afterthought is shown by the first lines of the ten books, which together give the acrostic GUILLERMUS, as contemporaries were quick to note. However, William was already William, and already a grand enough patron, when he was bishop of Chartres and also, from 1169, archbishop of Sens. In the poem there are just two references to him as archbishop of Rheims. First in the dedication itself, which is worth quoting in full (1. 12–26):

at tu, cui maior genuisse Britannia reges
gaudet avos, Senonum quo presule non minor urbi
nupsit honos quam cum Romam Senonensibus armis
fregit adepturus Tarpeiam Brennius arcem
si non exciret vigiles argenteus anser,
quo tandem regimen kathedrae Remensis adepto
duriciae nomen amisit bellica tellus,
quem partu effusum gremio suscepit alendum
phylosophia suo totumque Elycona propinans
doctrinae sacram patefecit pectoris aulam,
excoctumque diu studii fornace, fugata
rerum nube, dedit causas penetrare latentes:
huc ades et mecum pelago decurre patenti,
funde sacros fontes et crinibus imprime laurum
ascribique tibi nostram paciare camenam.

Here we have: (*a*) a prefatory line about William's royal descent; (*b*) nearly four colourful lines on him as archbishop of Sens, including Virgilian echoes, anti-Roman slant continuing that of the opening lines 1–11, and vindication of Brennus as a local anti-Roman hero;[15] (*c*) just two lines prosily recording his elevation to Rheims, decorated only by etymological allusion to its ancient name of Durocortorum; (*d*) five highly poetic lines on his devotion to learning, which lead naturally into the poet's request for support. In style and in construction the passage runs much better without the two lines about Rheims; and if William

[15] Contrast John of Salisbury, *Policraticus* 6. 7.

had already risen to be archbishop of Rheims when Walter first wrote the passage, the focus on his earlier post is hard to explain. Moreover, there is uncertainty in the MSS about where those two lines should go: some slot them in after line 23 or 24.[16]

The other reference to Rheims is also problematic. It occurs at the end of the peroration of book 5 (quoted above, p. 86–7), 518–20:

gens omnis et omnis
lingua Ihesum caneret et non invita subiret
sacrum sub sacro Remorum presule fontem.

Every people and every tongue would sing of Jesus and willingly submit to the holy font of the holy prelate of Rheims.

Presumably Walter imagined his patron accompanying his crusader king and baptizing infidels as he went along. But the lines risk conjuring up a vision of the entire Islamic world converging on Rheims that is little short of comic grotesque. I suspect that the last line was a belated and rather infelicitous substitute, born of a desire to salute the patron half way through the poem, once his elevation to Rheims allowed him a role in this context.

Apart from these two references to Rheims, the latest datable allusion in the poem is to the murder of Robert, archbishop of Cambrai in 1174. Of course the poem does not necessitate up-to-the-minute topical allusion. None the less, there is in fact enough of it to give substance to at least one argument from silence.

In November 1179 Walter's patron performed the office that the see of Rheims claimed as its special privilege: coronation of the king of France. The counter-claims of William's ex-see of Sens and the fact that the young king was his nephew gave the event a very personal significance, highlighted by Walter himself in a lyric poem.[17] The first book of the *Alexandreis* includes an elaborate account of the young Alexander's coronation at Corinth (1. 203–38). Corinth is described not only as the capital of Greece, but also as a city converted by St Paul; in any case Walter is not shy of modern comparisons (cf. especially 8. 168–71). If the coronation of a young king of France, 40 years since the last one and of particular significance to Walter's patron, had been a recent event, or was even on the cards, this passage was crying out for an allusion to it. There is none that I can see, not even implicit.

The same argument applies with greater force to the peroration of book 5 (510 ff.):

si gemitu commota pio votisque suorum
flebilibus divina *daret* clementia talem
Francorum regem, toto *radiaret* in orbe . . .

[16] Besides S as recorded by Colker, also at least London, BL Add. 23891, Novara Capit. N. 44, Oxford Bodl. 527; it should be noted that Colker's edition is based on a selection rather than a recension of the MSS.

[17] *Die Lieder Walters von Châtillon*, ed. K. Strecker (Berlin, 1925), poem 30.

If divine clemency, moved by the devout lament and tearful prayers of its people, granted such a king of France, then in the whole world would shine . . .

The subjunctives imply that divine clemency has not so far shown signs of obliging. If a young French king had just ascended the throne, even if he had not yet had time to prove himself an Alexander, it would not have required a sycophant to be just a little more optimistic. Indeed, the passage as it stands, with its emphasis on tears, could only be a senseless insult.

Yet it was also on the basis of this peroration that Christensen argued for the dating 1178–82 which has been generally accepted, though I do wonder how many have troubled to read the argument.[18] Briefly, it runs thus: (*a*) the final lines, 518–20, must refer to conversion of the Jews in France, since the archbishop of Rheims cannot be imagined baptizing the whole world; (*b*) the prayer in lines 510 ff. implies that Philip Augustus has not yet declared his crusade against the Jews, though it is hoped for by clerics like Walter who could have no other reason to wish for a bellicose king of France; (*c*) a crusade against the Jews was in fact declared in 1180, so this passage must have been written in the winter of 1179–80, and the rest of the poem either side of that date. But the fact is that neither in this passage, nor elsewhere in the poem, is there any suggestion of any crusade against the Jews;[19] and that, like it or not, it is 'gens omnis et omnis lingua' who will be baptized by the archbishop and these can only be the many peoples of Islam referred to in the preceding lines, which also amplify the prayer of 510 ff. The argument seems to me wholly unfounded.[20]

Moreover, there is independent evidence for dating the *Alexandreis*, which, instead of being deployed for this purpose, has itself been misinterpreted on the basis of the supposed date 1178–82. Among Walter's minor works there is a fascinating prosimetrum on the world of learning and its various branches. It presents itself as an address for the fourth Sunday in Lent to the assembled doctors and students of law at Bologna.[21] The first of the arts is introduced thus (7–8):

[18] Christensen (n. 1 above), 4–13.

[19] The Jews do figure in the poem, but as predecessors, not enemies, of Christianity, cf. 1. 420–1 'inde Palestinae cunctis supereminet una / unius Iudea Dei'. Alexander's dream and visit to Jerusalem (1. 499–554) are based on tradition, but the much-praised Jewish sculptor Apelles (Colker, *index nominum* s.v.) is Walter's creation and he includes Christian prophecy in his work (4. 256–67); the only note of criticism is 4. 242–5, which hardly suggests a crusade (cf. 2. 511–21). Walter's own dialogue *Adversus Iudaeos* is evangelizing, not bellicose.

[20] E. Herkenrath, 'Die Zeit der *Alexandreis* Walters von Châtillon', *Historische Vierteljahrsschrift*, 29 (1935), 597–8, dates the passage to the crusading fervour of 1188/9, which would be plausible if 510 ff. could possibly mean: 'Machte Gottes Gnade den König der Franzosen so erfolgreich wie Alexander . . .'; they cannot.

[21] *Moralisch-satirische Gedichte Walters von Châtillon*, ed. K. Strecker (Heidelberg, 1929), poem 3; cf. part 3: 'ante legum dominos et magistros artium / usurpare videor doctoris officium / ut sermonis epulo relevem ieiunium', and the MS rubrics: 'Galterus de Insula

inter artes igitur que dicuntur trivium
fundatrix grammatica vendicat principium.
sub hac chorus militat metrice scribentium;
inter quos sunt quatuor rithmice dictantium
qui super hoc retinent sibi privilegium,
Stephanus flos scilicet Aurelianensium
et Petrus, qui dicitur de castro Blesensium.

istis non inmerito Berterus adicitur,
set nec inter alios quartus pretermittitur
ille, quem Castellio latere non patitur,
in cuius opusculis Alexander legitur.

So among the arts that are called the trivium, grammar the foundress claims first place. Under her banner campaigns the chorus of metrical poets, among whom there are four who are also pre-eminent in rhyme, that is Stephen, the pride of Orleans, and Peter, known as from Blois; Berterus is added with good reason, nor among others is a fourth forgotten, the one whom Châtillon proclaims, among whose works Alexander is read.

The date of this prosimetrum can be fixed with fair precision within the years 1174–6.[22] Starting from the premiss that the *Alexandreis* was written later, Strecker was reduced to supposing that the last line referred to lost and otherwise unknown poems that Walter had written about Alexander, taking 'in cuius opusculis' to mean 'in several of whose small works', and suggesting that these works should belong to the class of 'rithmice dictata'. But the relative clause, like the preceding one 'quem Castellio . . .', serves to identify the poet; it does not need to refer to 'rhythmic' poems. Moreover, the diminutive is not to be pressed, since Walter in his own preface to the *Alexandreis* addresses 'lectores huius opusculi'; and the more natural sense of the line is that among his works there is one poem about Alexander, with the subject standing for the work as e.g. in Ovid, *Amores* 1. 15. 25 'Tityrus et segetes Aeneiaque arma legentur', or *Ars Amatoria* 3. 337–8 'et profugum Aenean, altae primordia Romae, / quo nullum Latio clarius exstat opus'. Above all, the line has no point unless it refers to something for which the poet was renowned; and it passes belief that the poet of the *Alexandreis* could have previously written famous poems on the same subject and no trace of

predicans scolaribus bon[oniensibus] in reditu suo a curia romana' and 'sermo recitatus Bononiae coram episcopo et scolaribus in dominica Letare Ierusalem'; since Bologna is not mentioned in the work, but Walter's presence there is independently attested in *Vita* 2 (Colker, p. xii), the rubrics seem trustworthy.

[22] Strecker, pp. 36–7; I see no reason to suppose that the list of poets was a later addition (cf. E. Herkenrath's review of Strecker's edition, *Historische Vierteljahrsschrift*, 26 (1931), 856–7). There is independent evidence that both Stephen of Orleans and Peter of Blois performed publicly at Bologna; nothing is known of Berterus. The choice makes much better sense as a list of precedents (French poets addressing the university of Bologna) than as a list of famous poets in general; indeed I wonder whether the rather obscure line 'qui super hoc retinent sibi privilegium' should not be translated 'who in addition can claim this privilege', viz. of having addressed the law faculty at Bologna.

them remain: it is just the sort of thing that accessus and scholia would have fastened on.

So I would conclude that the hypothesis is implausible and unnecessary, and that when Walter wrote these lines in 1174–6 he was already known to have written the *Alexandreis*.

But we have not yet finished with the prosimetrum. For further on, in part 14, we have twelve classical hexameter lines, the only ones in the work, capped by two rhyming ones, and the classical ones recur almost verbatim at *Alexandreis* 3. 140–57. Which is quoting which? Let us first look at the *Alexandreis* passage. In the thick of the battle of the Issus Alexander is faced with a warrior-astrologer who knows he is about to die and hopes to make it a glorious death by provoking Alexander (the parts recurring in the prosimetrum are in italics and its variants listed below):

stabat ab opposito niveis pretiosus in armis
Memphites Zoroas, quo nemo peritior astris
mundanas prenosse vices: *quo sydere frugis*
defectum patiatur ager, quis frugifer annus,
unde nives producat hyemps, que veris in ortu
temperies impregnet humum, cur ardeat estas,
quid dedit autumpno maturis cingier uvis,
circulus an possit quadrari, an musica formet
caelestes modulos, vel quanta proportio rerum
quatuor inter se. novit *quis sydera septem*
impetus oblique rapiat contraria mundo,
quot distent a se gradibus, que stella nocivum
inpediat sevire senem, quo sydere fiat
obice propitius, Martem quis temperet ignis,
quam sibi quisque domum querat, quod sydus in isto
regnet hemisperio. motus rimatur et horas
colligit, eventus hominum perpendit in astris.
Parva loquor, totum claudit sub pectore caelum.

142 esto secundus Atlas, dicas quo . . .
146 cur legat autumnus pregnantes ebrius uvas
148–9 c. m. dicas quis . . .
post 157 finge quod hec scieris, set et his maiora: quid ad rem? / tam cito descieris hec omnia, quod nihil ad rem.

Opposite him stood Zoroas of Memphis, sumptuous in snow-white armour, more skilled than any in foretelling the course of nature by the stars: by what star the crops would fail, what year would bring a full harvest, whence the winter produces snows, what warmth seeps into the soil at spring's rising, why the summer burns, what gave autumn a girdle of ripe grapes; whether the circle can be squared, whether music is a pattern of heavenly harmonies, or what is the proportion between the four elements of things. He knows what force rushes the seven planets cross-wise against the firmament, how many intervals there are between them, what star can prevent old man Saturn from raging, by what star's intervention he will become kind, what luminary can soften Mars, what home each planet seeks, what star rules in its hemisphere. He tracks each movement and reckons the hours, and weighs up the fortunes of men in the stars. I have said but a fraction: he encloses the whole sky in his mind.

The episode is invented by Walter, and it neatly balances the astrological episode at the end of the same book (501–25), which is freely elaborated from Curtius. For Zoroas, Walter found one starting-point in his favourite Lucan, where Nigidius Figulus is introduced as an astrologer (1. 639–41):

> at Figulus, cui cura deos secretaque caeli
> nosse fuit, quem non stellarum Aegyptia Memphis
> aequaret visu numerisque moventibus astra,

followed by a speech in which some of the themes of the Zoroas description appear in catastrophic tone (failure of crops, planetary oppositions). And the concluding idiom in line 157, 'Parva loquor', is also a tag from Lucan (9. 783). But verbally the beginning of the passage draws on a different epic model, the prayer to Apollo at the end of the first book of Statius' *Thebaid* (1. 705–6):

> tu doctus iniquas
> parcarum *praenosse* manus fatumque quod ultra est
> et summo placitura Iovi, *quis* leti*fer annus*,
> bella quibus populis, quae mutent sceptra cometae.

Walter's combination of *praenosse* and *quis* ¯˘*fer annus*, both in the same metrical position and with neat inversion of sense in the second, can hardly be coincidental. But this shows that lines 142–3 in the *Alexandreis* passage were composed as a unit; the split found in the prosimetrum must be secondary.

This conclusion would in any case follow from comparing the functions of the two passages in context. In the *Alexandreis* the emphasis on astrology, confirmed in the concluding epigram, 'totum claudit sub pectore caelum', is integral to the scene.[23] In the prosimetrum, the purpose of the passage is to take leave of the arts of the quadrivium, rather summarily treated in part 11. There was just enough general science in the Zoroas description to warrant its re-use in this sense, but even after the omissions the emphasis on astronomy/astrology is rather off key—contrast the careful balance in the parallel dismissal of the trivium that precedes. But on a literary level there is ample compensation for the imbalance if we read the passage as an elegantly adapted self-quotation, confirming with gentle irony the previous self-advertisement. The jingling couplet added at the end:

> finge quod hec scieris, sed et his maiora: quid ad rem?
> tam scito descieris hec omnia, quod nihil ad rem

[23] Manilius 2. 121 is curiously similar; it would be interesting to see whether there are any other parallels with this rare poet.

gives a smooth transition back to medieval measures just at the point where the *Alexandreis* description switches back to narrative. On the opposite view, that the prosimetrum version was composed first, its combination of epic style and echoes, astrological emphasis and ironic flourishes seems to me hard to explain.[24]

I conclude then that the *Alexandreis* was largely complete before 1176 and probably published in that year with a couple of last-minute additions about Rheims. It was as fine a promotion-present as any patron could expect from his poet. But does it matter whether the *Alexandreis* was finished by 1176 or not till 1182/4? I think it does, and not just for the sake of getting details right rather than wrong. The *Alexandreis* is a historical poem on the most public and political of themes: conquest and rule. Properly to understand it will require more knowledge than I have of the political history and thought of the period. But it may be a start to determine just what period we are talking about.

[24] I would add that the prosimetrum version of l. 146 is a decided improvement on the other, completing the personification of Autumn as a tipsy *vindemiator*. Strecker's argument from the variant reading *eclipsim* for *defectum* in l. 143 is invalid because the same variant is found in *Alexandreis* MSS and may well be a gloss.

6 Classical Latin Satire and Medieval Elegiac Comedy

GIOVANNI ORLANDI

The history of satire in the western Middle Ages is a very complicated one. Continuity between ancient Roman and medieval satiric poetry is generally difficult to perceive; authors such as Amarcius and, to a lesser degree, Marbod, whose imitation of Horace is conscious and obvious, are rare exceptions.[1] True, in twelfth-century Europe there are great poets who sometimes explicitly described their work as *satura*—Walter of Châtillon is a good example. Yet he, and others besides him, selected for their satiric pieces a form radically different from that employed in classical satire: the rhythmical, rhymed strophes of liturgical hymns and sequences, whose language and style were influenced much more by the Psalms and prophets than by Roman satirists. Thus, at first glance, Latin satire of the twelfth century would seem to owe nothing to its ancient predecessors; or perhaps one should say, such debts as it has, it successfully conceals.

Take, for example, the Archpoet's excuse for not being able to compose an epic celebrating Barbarossa: he mentions practical obstacles which prevent him from writing and asks for material benefits which will help him to do so: a quiet and comfortable life, plenty of food, and, above all, good wine. Here the possible source of inspiration may be seen in the beginning of Horace's nineteenth *Epistle* (2–34): 'nulla placere diu nec vivere carmina possunt / quae scribuntur aquae potoribus' ('no poems written by water-drinkers can please or last for long').[2] Another—and in my view,

I am grateful to Professors Jill Mann and Michael Lapidge for their unfailing help and advice.

[1] See Sextus Amarcius, *Sermones*, ed. K. Manitius (Weimar, 1969), 20, 26; Marbodus, *Liber Decem Capitulorum*, ed. R. Leotta (Rome, 1984), 59–72, listing parallels from Horace's *Ars poetica*.

[2] Cf. *Die Gedichte des Archipoeta*, ed. H. Watenphul and H. Krefeld (Heidelberg, 1958), 57–61 (poem 4). For this and other parallels see P. Dronke, above, ch. 4. I am obliged to Professor Dronke for some valuable suggestions.

more probable—source is *Epistles* 2. 2. 77–80, where we find the *scriptorum chorus* echoed by the Archpoet: 'ieiunant et abstinent poetarum chori' (11. 1); one may also compare 'vis canere' in Horace (80) with '*Vis* ut infra circulum parve septimane / bella *scribam* forcia' in the Archpoet (5. 1–2).

Even more obvious is the influence of Juvenal on a passage from a satiric song by Walter of Châtillon:[3]

Cardinales, ut predixi,
novo iure crucifixi
vendunt patrimonium;
Petrus foris, intus Nero,
intus lupi, foris vero
sicut agni ovium.[4]

Tales regunt Petri navem,
tales habent eius clavem,
ligandi potentiam;[5]
hi nos docent, set indocti,
hi nos docent et nox nocti
indicat scientiam.[6]

In galea sedet una
mundi lues inportuna
camelos deglutiens;[7]
involuta canopeo
cuncta vorat sicut leo
rapiens et rugiens.[8]

Hic piratis principatur
et Pilatus appellatur,
sedens in insidiis;[9]
ventre grosso, *lota cute*[10]
grande monstrum nec virtute
redemptum a vitiis.[11]

The cardinals, as I said before, according to a new law, sell Christ's patrimony; their appearance is that of Peter, their inner self that of Nero; inwardly they are wolves, outwardly little lambs—such are they who steer the ship of Peter, such are they who hold his keys, the power

[3] *Moralisch-satirische Gedichte Walters von Châtillon*, ed. K. Strecker (Heidelberg, 1929), 26–7 (poem 2). The fact that Walter would term this and other similar poems *saturae* (cf. ibid., 46 (3. 23. 3), 124 (13. 3. 4), 126 (13. 15. 4), 134 (15. 2. 6)) has been pointed out to me by Professor C. J. Classen.

[4] Cf. Matt. 7: 15; Psalm 113: 4, 6.

[5] Cf. Matt. 16: 19.

[6] Cf. Psalm 18: 3.

[7] Cf. Matt. 23: 24.

[8] Cf. Psalm 21: 14.

[9] Cf. Psalm 9: 28.

[10] Cf. Juvenal 6. 464.

[11] Cf. Juvenal 4. 2–3 'monstrum nulla virtute redemptum / a vitiis'. All references are taken from Strecker's commentary.

of binding; these are the ones who teach us, though they know nothing, these are our teachers, and night mirrors its dark knowledge to night. On the galley sits the heinous pest gobbling up camels, covered with a canopy, devouring everything like a lion, preying on it and roaring. He is the chief of pirates—called Pilate, sitting in ambush—large-bellied, well-washed, a huge monster—unredeemed by any virtue from his vices.

The general tone of the poem is determined by the accumulation of biblical expressions and shaped by very unclassical verses and rhymes. In this context Strecker identified a passage from Juvenal (reproduced almost verbatim, with the addition of *grande* and the substitution of *nec* for *nulla*), preceded by another phrase borrowed from the same author (*lota cute*, 'with pampered flesh'). Thus Walter's ability—pointed out by Christensen with reference to the *Alexandreis*—to turn Curtius Rufus' prose into flawless hexameters with the minimum of change,[12] is here employed to the opposite end of converting the classical verses of Latin satire into the rhythms of twelfth-century songs. The final outcome is something very far from the atmosphere of ancient Latin satire, in spite of the great number of classical lines, mainly from Juvenal, employed as concluding verses in the Goliardic strophes *cum auctoritate*.[13] In short, even if first-hand knowledge of classical satire in the twelfth century may be taken for granted,[14] neither the language nor the structure of Goliardic satiric poetry seems to have been affected by it.

I think it more interesting, therefore, to investigate how the links between comedy and satire, as set out in the theoretical writings of authors such as Isidore,[15] are realized in the literary practice of the Latin Middle Ages. Suchomski and Kindermann have shown that these two genres are customarily linked with

[12] H. Christensen, *Das Alexanderlied Walters von Châtillon* (Halle, 1905), 102 ff.

[13] See e.g. Strecker (n. 3 above), 63–70 (poem 4), the references ibid. 71, and P. G. Schmidt, above, 39–55.

[14] This must be true at least for Walter, and for other poets as well. Yet many contemporary scholars have surely contented themselves with using collections of excerpts; see in general B. Munk Olsen, 'Les classiques latins dans les florilèges médiévaux antérieurs au XIIIe siècle', *Revue d'histoire des textes*, 10 (1980), 47–172, and, regarding Orléans in particular, R. H. Rouse, 'Florilegia and Latin Classical Authors in Twelfth- and Thirteenth-Century Orléans', *Viator*, 10 (1979), 131–60 (especially concerning the well-known *Florilegium Gallicum*, preserved in a dozen manuscripts and going back to the middle of the twelfth century).

[15] *Etymologiarum sive Originum libri XX*, ed. W. M. Lindsay (Oxford, 1911), 8. 7. 7 'Duo sunt autem genera comicorum, id est, veteres et novi. Veteres, qui et ioco ridiculares extiterunt, ut Plautus, Accius, Terentius. Novi, qui et Satirici, a quibus generaliter vitia carpuntur, ut Flaccus, Persius, Iuvenalis vel alii'.

each other in the rhetorical treatises of the twelfth and thirteenth centuries, and I take their observations as the starting-point of my own enquiries.[16] In the present paper, I shall try to give a preliminary and provisional answer to the question how far particular classical satires influenced a literary genre in which comic and satiric intentions are intertwined—namely, the so-called elegiac comedy.

The influence of ancient Latin poetry on language and style has been noted by the editors of almost every comedy;[17] yet classical satirists figure little if at all in their lists of parallels, especially in comparison with other poets such as Ovid, whose profound influence on Latin twelfth-century poetry is beyond doubt. Even allowing for the fact that scholarly attention has been focused on Ovid and Virgil from the outset, so that research on other authors has been relatively neglected, so striking a disproportion must be accounted for. One would expect, for instance, that the celebrated, or notorious, description of the visit of Spurius to Spurca in William of Blois's *Alda* would be filled with reminiscences of ancient satire; yet such reminiscences as I can find add up to almost nothing.[18] To cite one case: one can only wonder whether it is by chance that within two lines (*Alda* 310–11 'Spurca domi sola residet villosaque crudo / cum sale *pannosi suminis* exta vorat, / et fragmenta bibit in *pultes* iam resoluta / hesterni iuris atque aliunde dati') there are three words which also occur in Persius: *pannosus* (4. 32), *sumen* (1. 53), *puls* (6. 40). Yet no firm inference can be drawn from this. Or again, in the line (*Alda* 558) 'infundens *bibulis auribus* urbis ea', one might suspect an echo of Persius 4. 50, 'nequiquam populo *bibulas* donaveris *aures*'; however, the possibility of an intermediary source cannot be ruled out. Likewise, the combination *murmura rodit* used by the anonymous

[16] See J. Suchomski, '*Delectatio*' *und* '*utilitas*': *Ein Beitrag zum Verständnis mittelalterlicher komischer Literatur* (Berne and Munich, 1975), 142 ff.; U. Kindermann, '*Satyra*': *Die Theorie der Satire im Mittellateinischen*: *Vorstudie zu einer Gattungsgeschichte* (Nuremberg, 1978), 144 ff.

[17] Practically no reference is given by individual editors of the standard corpus of elegiac comedies produced under the direction of G. Cohen, *La* '*comédie*' *latine en France au XII*^{e} *siècle* (Paris, 1931). The results of previous research in this field (e.g. H. Hagendahl, 'La 'comédie' latine au XIIe siècle et ses modèles antiques', in *Δράγμα Martino P. Nilsson . . . dedicatum*, ed. K. Hanell, E. J. Knudtzon, N. Valmin (Lund, 1939), 222–55), together with new discoveries, are conveniently assembled by the editors of the new corpus, under the direction of F. Bertini, *Commedie latine del XII e XIII secolo* (Genoa, 1976–).

[18] We still have to quote from the edition by M. Wintzweiler, in Cohen (n. 17 above), i. 130–51, reprinted in *Lateinische Comediae des 12. Jahrhunderts*, ed. and tr. J. Suchomski and M. Willumat (Darmstadt, 1979), 206–34. A new edition by F. Bertini is being prepared for vol. 6 of his corpus (see above, n. 17).

author of *Baucis et Traso* (31) is found only in Persius 3. 81 ('*murmura* [. . .] et rabiosa silentia *rodunt*');[19] yet it is the only parallel with Persius to be found in the *Baucis*, and before assuming direct dependency it would be wise to wait for more indexes and concordances to poets of the period between late antiquity and the twelfth century to become available.

Thus unquestionable verbal echoes, so often identified in Ovid and, for some texts, in Terence,[20] seem relatively rare in the case of the ancient satirists. One possible explanation for this dearth of reminiscences is the intrinsic difficulty of reading and understanding poetry of this sort, especially Persius and Juvenal. Horace would have been easier, of course, but was, on the other hand, the least satiric of the three from a medieval point of view—that is, the least explicitly moralizing. The consequences of this fact have been illustrated by Bischoff in a suggestive essay: whatever Horace wrote, even the *Odes*, were 'misinterpreted in a most incredible manner in order to make them fit this medieval concept' (the role of the satirists 'as censors and preachers of morality').[21]

Horace is in any case, so far as I can judge, more frequently echoed in elegiac comedy than the other two writers. For example, a passage of Horace, in my opinion, lies behind a short poem of Hildebert of Lavardin, which the main manuscripts preserve among a series of epigrams but which I think is to be linked to elegiac comedy. It may, indeed, be the *argumentum* of a lost elegiac comedy, or of one that was planned but not realized:[22]

> Servus ait domino 'gratis famulabor in annum,
>   si facis ut famula qualibet arte fruar'.
> poscit herus coitum, sibi ponit servula noctem.
>   ille thoro famulum collocat, hec dominam.

[19] See my edition in Bertini (n. 17 above), iii. 270–302.

[20] S. Pittaluga, 'Echi terenziani nel *Pamphilus*', *Studi medievali*, Ser. 3, 23 (1982), 297–302. Not entirely convincing is the discussion of A. K. Bate, 'Language for School and Court: Comedy in *Geta*, *Alda* and *Babio*', in *L'eredità classica nel medioevo*: *Il linguaggio comico* (Viterbo, 1979), 145.

[21] B. Bischoff, 'Living with the Satirists', in *Classical Influences on European Culture*, A.D. *500–1500*, ed. R. R. Bolgar (Cambridge, 1971), 90 = *Mittelalterliche Studien*, iii (Stuttgart, 1981), 267.

[22] Hildebertus Cenomannensis, *Carmina minora*, ed. A. B. Scott (Leipzig, 1969), 16 (poem 24). I am preparing a paper discussing the motifs of this 'epigram' in connection with various vernacular novels. Another link between Hildebert's epigrammatic poetry and elegiac comedy may be traced in two short poems describing the merchant Milo with the typical features of the cuckold (ibid. 3, poems 7, 9). See below, n. 40.

res patuit casu, vir nuptam *pro*vocat *in ius*.
facta refert; *risum* facta relata movent.
uxor et ancilla *laudantur*, virque monetur,
ne messem famuli sic emat ulterius.

A servant said to his lord: 'I will work a year for nothing, if you will enable me to enjoy a maid-servant by whatever means you like.' When the master asked for sex, the little maid-servant fixed a time for a night-assignation. He put his servant in the bed, she her mistress. The affair was revealed by chance: the husband took his wife to court; related what had happened; and that made everyone laugh. The wife and the maid were praised; and the husband warned not to pay for the servant's harvest in that manner in the future.

If one bears in mind that Hildebert died in 1133, this means that the piece is very near in date to the origins of the genre.[23] The story of Hildebert's poem is a typical example of a *comedia*-plot. A servant asks his master to procure for him the sexual favours of the serving-maid; the master asks the maid to sleep with *him*;[24] the maid fixes the night of their meeting, but persuades her mistress to take her place; meanwhile the servant, according to plan, takes his master's place, and so the master contrives his own cuckoldry. This piece is eight lines long, just like the *argumentum* of the *Alda*, whereas those of Vitalis of Blois are slightly longer: ten lines each. Some elements in the final trial-scene are probably reminiscent of a similar episode in Horace, *Satires* 1. 7. 20–3: '*in ius* / acres *pro*currunt, magnum spectaculum uterque. / Persius *exponit causam. ridetur* ab omni / conventu. *laudat* Brutum laudatque cohortem', etc. The cluster of reminiscences of Horace (the phrase *in ius*; the audience laughing; the verb *laudo*, though used differently) seems to guarantee the later poet's dependence; for Hildebert's direct knowledge of Horace is beyond doubt.[25]

A similar situation obtains with respect to a passage of *Baucis et Traso* whose dependence on Juvenal escaped me when I was editing its text:[26]

[23] One has to bear in mind that a number of elegiac comedies must have been lost, as we know from the *Registrum* of Hugo of Trimberg, who, writing in the second half of the 13th cent., mentions five comedies which were—he says—read in school, and whose *incipits* he quotes. Two of them have not survived, and one, *Lidia*, is preserved in only two copies. Cf. K. Langosch, *Das* Registrum Multorum Auctorum *des Hugo von Trimberg* (Berlin, 1942), 191, 251 (ll. 716–19).

[24] I have argued elsewhere in favour of placing the comma after, and not before, *sibi* (l. 3): see *Studi medievali*, ser. 3, 15 (1974), 1048.

[25] Cf. the apparatus of *loci paralleli* in Scott (n. 22 above), and *Studi medievali* (n. 24 above), 1026.

[26] Cf. Bertini (n. 17 above), iii. 282–4 (ll. 109–14).

Traso domum rediens compensat opus famulorum:
hec probat, hec culpat, hec male gesta refert;
hos scutica cedit, hos punit pena flagelli,
verberat hos virgis, hos trahit unca manus.
dampna luunt domini famulorum sedula turba,
et sic alterius crimine punit eos.

Returning home, Thraso rewards the work of his servants, commending one, scolding another, saying that certain things were ill done, scourging some, whipping others as punishment, lashing some with thongs and dragging others along with bent fingers. The diligent throng of servants are punished for their master's misfortunes and suffer for the wrongs done by others.

At that time I was satisfied with Horace (*Satires* 1. 3. 119 'ne *scutica* dignum horribili sectere *flagello*'); yet a comparison with Juvenal 6. 475 ff. makes it immediately clear that the latter is the source, and not, or not only, Horace:

si nocte maritus
aversus iacuit, periit libraria, ponunt
cosmetae tunicas, tarde venisse Liburnus
dicitur et *poenas* alieni pendere somni
cogitur, *hic* frangit ferulas, rubet ille *flagello*,
hic scutica; sunt quae tortoribus annua praestent.
verberat atque obiter faciem linit, audit amicas
aut latum pictae vestis considerat aurum
et *caedit*, longi relegit transversa diurni
et *caedit* . . .

If her husband, the night before, has slept with his back to her, then the wool-maid's had it, the dressers are stripped and flogged, the litter-bearer is accused of coming late, and is compelled to pay the punishment for another man's sleep. One victim has rods broken over his back, another bears bloody stripes from the whip: another is lashed with a cat-o'-nine-tails: some women pay their floggers an annual salary. While the punishment's carried out she'll be fixing her face, gossiping with her friends, giving expert consideration to the width of the hem on some gold-embroidered robe and as the flogging goes on she skims through the daily gazette.

Apart from single parallels—not only *flagellum* and *scutica* but *poena* and *verbero*—it is the content of this scene which provides the decisive clue; in both passages the wretched servants pay for the faults of others, the Roman matron punishing them because she had failed to attract her husband the previous night, and Traso because he could not yet arrange an appointment with his beloved

Glicerium. The author of *Baucis et Traso* makes more explicit the meaning of Juvenal's passage in a verse which is a remarkable *sententia* ('dampna luunt domini . . .') and which, I am now inclined to think, would be better punctuated if it were isolated by means of brackets, in order not to break the continuity between lines 112 and 114. As to *famulorum turba*, I should here even surmise the influence of Seneca's famous *Epistle* 6. 47 with its very sympathetic treatment of slaves' status, where the verbal combination *servorum turba* is found.[27]

Elsewhere the links are more difficult to demonstrate, and require more thorough scrutiny. I shall try to comment upon some passages in Horace, *Satires* 1. 2, which in my opinion supplied medieval authors with particular words and, on occasion, with thematic material for the action of their plots. Geta's self-portrait in Vitalis of Blois's first drama involves a detailed description of each of his limbs, where not even the penis is passed over:[28]

> Sic tumeo ventre quod dicor ydropicus esse
> et stomacus parvum nescit habere modum
> zonaque nulla potest hunc castigare tumorem
> cum pane absortis dum tumet alvus aquis:
> deest spatium lateri, deest renibus. hispida crura
> sunt michi que scabies ut sua regna tenet.
> Sed michi dum crebro singultu colligit *iram*,
> ad curvum *muto* tenditur usque genu.
> tibia curta michi sed grossa pedesque recurvi
> ut pedibus solea nulla sit apta meis.

So does my belly swell that I am held to be dropsical, and my stomach knows no limits; no girdle can contain my swelling as my belly expands

[27] *Epistle* 6. 47. 2. In passing, one may suggest that some words of the same passage, if they were punctuated and understood without inverted commas ('Servi sunt, immo homines; servi sunt, immo contubernales; servi sunt, immo humiles amici; servi sunt, immo conservi'), might have influenced Héloïse's epistolary style: 'Quod itaque sorori immo ancillae concessistis, frater immo dominus impleatis' (letter to Peter the Venerable, ed. G. Constable, *The Letters of Peter the Venerable* (Cambridge, Mass., 1967), i. 400–1 (*Ep.* 167)). I wonder whether this passage has yet been adduced in the course of the endless controversy about the authenticity of Heloïse's letters to Abelard, one of which bears the celebrated *inscriptio*: 'Domino suo immo patri, coniugi suo immo fratri, ancilla sua immo filia, ipsius uxor immo soror, Abelardo Heloysa' (cf. Abélard, *Historia calamitatum*, ed. J. Monfrin (Paris, 1959), 111). Here one may also trace the possible influence of a passage of St Jerome's letter to Eustochium (*Ep.* 22. 26): 'mi Eustochia, filia, domina, conserva, germana'.

[28] Bertini (n. 17 above), iii. 182–241 (ll. 341–50). In Bertini's apparatus one misses the reference to Persius (5. 132–3) which may explain the origin of a well-known (and imitated) passage of the *Geta* (61–8): 'Cito, Birria, surge', etc. Professor Bertini generously draws my attention to two interesting parallels from Latin satirists to the *Lidia*, l. 97 (cf. Juvenal 4. 3) and l. 187 (Horace, *Epodes* 7. 15).

with the water that I've drunk with bread: there is no room for my side, no room for my kidneys. My hairy legs are covered with a scurf which holds them in thrall. But when my penis begins to grow angry with sigh upon sigh it hangs down to my knobbly knees. My shins are short but fat, my feet turn back on one another, so that no shoe will fit my feet.

Now the term *mut*(*t*)*o*, 'penis', is to be found only in Horace (*Sat.* 1. 2. 68, a passage explained by Porphyrio and ps.-Acro as deriving from Lucilius):[29]

> huic si *muttonis* verbis mala tanta videnti
> diceret haec animus: 'quid vis tibi? numquid ego a te
> magno prognatum deposco consule cunnum
> velatumque stola, mea cum conferbuit *ira*?'

If while facing such evils a man's mind were thus to plead on his penis' behalf, 'What do you want? Do I ever, even when my desire is raging, ask you for a cunt clad in a *stola*, descended from a great consul?'

Here the relevant context is so radically different from that of the *Geta* that one would be inclined to dismiss the possibility of a direct relation (for *muto* is also found in glossaries),[30] were it not for the presence of *ira* in both passages with the particular meaning of 'sexual excitement', which suggests direct dependence (whereas 'iram colligit' in Horace, *Ars poetica* 160, has nothing to do with this subject and therefore may be a mere verbal echo). The use of *singultus* may come from Persius 6. 72, 'cum morosa vago *singultiet inguine* vena', especially since the meaning of *inguen* is thoroughly exploited in a very similar passage just 15 lines later (*Geta* 363–6).[31]

It can be proved that one single word of the same *Satire* of Horace, understood in a particular way, gave rise to a whole scene of *Baucis et Traso*. The knight Traso wants to go to bed with Glicerium, a supposed virgin lodging at Baucis' house (in fact she is a whore disguised as a virgin by the smart *lena*). Awaiting an appointment with Baucis, he hides in a ditch, but cannot escape Birria's attention. Birria, a base servant, at first thinks he will drive him out by throwing stones at him,[32] but soon afterwards

[29] Lucil. fr. 307 Marx (= 308 Krenkel, 320 Terzaghi-Mariotti). Cf. I. Mariotti, *Studi luciliani* (Florence, 1960), 106.

[30] G. Loewe, *Prodromus corporis glossariorum Latinorum* (Leipzig, 1876), 304: 'muto: priapus'. Cf. *Corpus glossariorum Latinorum*, ed. G. Goetz, vi (Leipzig, 1899), 722.

[31] Cf. *Alda* 467–8, probably influenced by the *Geta*.

[32] This may well be an allusion to a notorious scene in the *Geta* (139–212).

he has a new idea: 'permingatur', he says, 'sufficit iste pudor' (240), and poor Traso is instantly soaked with urine. The source of this development of the plot may well be some current story, although I do not know any extant *fabliau* which corresponds in this respect to our comedy. However, the verb *permingo*[33] is rare enough to encourage another hypothesis (which of course does not exclude the previous one). In a well-known passage of *Satire* 1. 2 (41–6) Horace lists the disgraceful accidents which usually afflict adulterers:

> hic se praecipitem tecto dedit; ille flagellis
> ad mortem caesus; fugiens hic decidit acrem
> praedonum in turbam; dedit hic pro corpore nummos;
> hunc *perminxerunt* calones; quin etiam illud
> accidit, ut cuidam testis caudamque salacem
> demeterent ferro.

One man has thrown himself headlong from the roof; another has been flogged to death; a third, in his flight, has fallen into a savage gang of robbers; one man redeemed his life with cash; another has been buggered by stable-boys. Once it so happened that a man had his testicles and lusty member cut with a sword.

As to *perminxerunt calones*[34] (which occurs uniquely here), we know for certain what it means in this context: as J. N. Adams has brilliantly shown, the adulterer was occasionally sodomized by the husband himself or, as here, by some of his slaves.[35] The same meaning of *meio* (= *mingo*) is found in *Satires* 2. 7. 52—a passage familiar to the authors of these comedies, as we soon shall see. But a medieval reader who had not had the chance to consult Porphyrio's commentary could only have understood *permingere* (or *permeiere*) literally, that is, as meaning 'to urinate on/at'

[33] The original form of the verb is *permeio* (and likewise, *meio* and *immeio*). Yet the forms *mingo, immingo, permingo*, which originated in late antiquity, seem to be the only ones known by medieval authors.

[34] Various scholiasts explain *calones* in this passage as *servi, asinarii, viles operarii* (cf. the scholia ed. H. J. Botschuyver (Amsterdam, 1935–42), i. 266–7, iv. 198), whereas ps.-Acro says 'calones sunt ministri militum, liberi homines' (ed. O. Keller (Leipzig, 1902–4), ii. 21). These statements were not considered contradictory in the twelfth century, when servants were no longer slaves. Thus the fact that the punishment of Traso is executed by Birria, 'servorum fex pessima' (215) probably confirms this passage of Horace as a source of *Baucis et Traso*.

[35] J. N. Adams, *The Latin Sexual Vocabulary* (London, 1982), 142. The correct interpretation of Horace's words is to be found in Porphyrio ad loc., ed. A. Holder (Innsbruck, 1894); but Porphyrio was rare in the Middle Ages, 'rarissimo' indeed (C. Villa, *La* lectura Terentii, i: *da Ildemaro a Francesco Petrarca* (Padua, 1984), 20). Cf. D. Fehling, *Ethologische Überlegungen auf dem Gebiet der Altertumskunde: Phallische Demonstration—Fernsicht—Steinigung* (Munich, 1974), 21 n. 90.

someone.[36] In other words, the episode of Birria humiliating Traso probably has its origin in a misinterpretation of a passage of Horace.

In the *Babio*, on the other hand, the punishment belongs to the same series given in Horace: *testes demetere* (45–6), an action well stressed by various scholiasts.[37] Babio is eventually castrated after trying in vain to deflower his stepdaughter Viola, thereby betraying his wife Petula, while he is in fact betrayed by Petula and Fodius.[38] He thus typifies the adulterer (even if a failed one), whereas Traso was merely a fornicator. But there was nothing to hinder a medieval *clericus* from confusing both categories of sinners within the larger classication of *moechi*, and indeed another passage of Horace may have encouraged such a confusion: 'te coniunx aliena capit, meretricula Davum: / peccat uter ostrum cruce dignius?' (*Sat*. 2. 7. 46–7). The comparison is clearly meant to favour the latter (the speaker is Davus himself), as the following lines show; yet one could conclude that, in varying degrees, both were *cruce digni*, and therefore worthy of an appropriate punishment. And indeed, although both Traso and Babio may well be regarded as victims of other people's sadism, no hint of solidarity with, or pity for, them is to be found in these texts; they both personify the vice of lust according to the technique of *exempla*.[39]

In another comedy a passage from the same *Satire* probably shaped, or helped to shape, the central conception and structure of the whole text; this, in my opinion, is the case with Matthew of Vendôme's *Milo*.[40] Here we are lucky enough to be able to identify various texts which closely resemble the probable source of this comedy's plot. In the *incipit* and *explicit* of his play Matthew alluded to a Greek tale which was allegedly the *materia* of *Milo*, a

36 Cf. ps.-Acro ad loc. (n. 34 above). The literal explanation of *perminxerunt* has wrongly been repeated by some modern scholars; cf. C. Sittl, *Die Gebärde der Griechen und Römer* (Leipzig, 1890), 100.

37 e.g. schol. *φψ ad. Ep*. 1. 2. 43 'dedit hic pro corpore nummos: Ne testones ei abscinderentur, poenam pecunia redemit' (Botschuyver (n. 34 above), i. 266); cf. ps.-Acro ad loc. (n. 34 above) 'Non, ne stupraretur, sed melius, ne castraretur'. Husbands castrating their wives' lovers when they caught them in the conjugal bed may also be found in *fabliaux*; cf. Ph. Ménard, *Les fabliaux, contes à rire du Moyen Age* (Paris, 1983), 21.

38 Cf. the edition by A. Dessì Fulgheri, in Bertini (n. 17 above), ii. 244–301.

39 Cf. J. Suchomski, 'Transformationen des komischen Erbes in der *nova comedia*: Komödie oder Satire?', in *L'eredità classica* (n. 20 above), 115–16.

40 Ed. F. Munari, *Mathei Vindocinensis opera*, ii (Rome, 1982), 59–72. Cf. Matteo di Vendôme, *De Afra et Milone*, ed. P. Busdraghi, in Bertini (n. 17 above), i. 168–95. One of the two MSS preserving the text interpolates Hildebert's Epigram 7 about the merchant Milo (n. 22 above) after l. 68; cf. Busdraghi, 178.

tale told in Constantinople (5–6, 253–4): it is the story of the farmer and the lion's footprints, preserved in different versions as one of the *Tales of Sindbad* or *The Seven Wise Masters*. One of these versions is in fact Byzantine, and (together with a Syriac version)[41] it can provide a clear picture of what one branch of the original tradition of the tale looked like: I am speaking of the *Syntipas* of Michael Andreopulos, writing at the end of the eleventh century. Let me give a sketch of the story in its (probably) original form:[42]

A king one day saw a woman, and instantly coveted her. He then sent her husband far away on an errand, went to her house, and asked her to yield to his desire. She answered: 'My lord, I am your servant, do what you wish to me; but first read this book.' The book condemned the injustice of adultery. When the king read it, he quickly got up and left; but his ring had fallen under the sofa. After the woman had got rid of him, her husband came back, and, sitting on the sofa, saw the ring. He recognized it as the king's ring, and thought that the king had dishonoured him with his wife; but, as he was afraid of the king, he refrained from any intercourse with her for a long time. The woman therefore told her father and brothers that her husband had become estranged from her. They went before the king and said: 'We had a field and committed it to this man so that he might work it; formerly he did so, but now he does not look after it and the field is lying waste.' The king asked the woman's husband: 'What have you to say?' He answered: 'My lord, it is true, I once worked the field as much as I could, but afterwards I saw a lion's footprints on it and for fear of the lion I worked there no more.' The king replied: 'In fact the lion passed by the field, but he did no harm; go back to your field and cultivate it in peace.'

In Matthew of Vendôme's text the husband is Milo, a peasant working hard all day long in the fields, while Afra, his wife, soon yields to the king's courtship—indeed she falls in love with him and begins to despise her husband. Milo begins to suspect what is happening, but his love dispels such suspicions, until one day he comes back from work at an unusual time and finds the door of the house locked. Inside the house is the king, who is compelled to flee in haste without putting on his distinctive regal sandals. Finally Milo, at the height of his rage, manages to break into the house; he finds the king's sandals in the bedroom and is convinced both of

[41] I use two modern translations of the latter: F. Baethgen, *Sindban, oder Die Sieben Weisen Meister: Syrisch und Deutsch* (Leipzig, 1879), 14; F. Macler, *Contes syriaques: Histoire de Sindban mise en français* (Paris, 1903), 20–2.

[42] Michael Andreopulos, *Liber Syntipae*, ed. V. Jernstedt = *Mémoires de l'Académie Impériale des Sciences de St.-Pétersbourg*, ser. 8, Cl. Hist.-philol., 11. 1 (1912), 13–15.

the adultery and the identity of the adulterer. Henceforth his jealousy forbids him to have either sexual or social relationships with Afra, who for her part feels so humiliated that she turns to her brothers and complains of being neglected by Milo. The remainder of the plot corresponds more or less to the oriental tale.[43]

A comparison of the two versions of the story reveals at once some striking contrasts:

(1) In the oriental collection the woman is a model of the wise wife, who even makes a despot repent of what he is about to do; on the contrary, Milo's wife Afra, though she is at first compelled by the king apparently against her will, is on the whole a typical example of the *moecha*, with all the misogynist attributes to be found in elegiac comedies, such as lust, ambition, eagerness for evil acts, etc.[44]

(2) The king in the Byzantine novel wields an absolute power, so that only an inner moral constraint, stimulated by reading the book, prevents him from satisfying his longing for the woman; the king described by Matthew, on the other hand, has his power undermined because of the illicit act he is committing, so he acts like a thief; he, the supreme authority, flees at the arrival of a man of the lowest station.

(3) Whereas no adultery has been committed in the oriental source, in the *Milo* the affair between Afra and the king is a *donnée* from which the plot develops, and which has noteworthy consequences in the final scene of the judgement. Here the main offender sits as the judge—a situation whose monstrosity is stressed from the outset by the author with sharp sarcasm (73–6), and alluded to by Milo himself in the final speech in the king's presence (195–6): 'Iuris ad articulum ritus regalis anhelet, / quo duce, quo racio preside tuta sedet.'

As far as point (1) is concerned, there is obviously no need to go back to Juvenal's sixth *Satire* to find a source for Matthew's

[43] The fundamental features characterizing the Syriac and the Byzantine tale are shared by other (e.g. Persian, Hebrew, and Spanish) versions: cf. *The Book of Sindibād: From the Persian and Arabic*, tr. W. A. Clouston (Glasgow, 1884), 80–3; *Tales of Sendebar*, ed. and tr. M. Epstein (Philadelphia, 1967), 92–101; *El libro de los engaños*, ed. J. E. Keller (Chapel Hill, 1953), 7–8, or *Libro de los engaños e los asayamientos de las mugeres*, ed. E. Vuolo (Naples, 1980), 13–17. In all these texts the husband's behaviour is determined by fear of the king. Later Arabic redactions introduce the husband's jealousy (cf. J. Scott, *Tales, Anecdotes, and Letters translated from the Arabic and Persian* (Shrewsbury, 1800), 69–75), but this shared motif is not sufficient to establish a link with the *Milo*.

[44] A general picture of this feature of the genre may be found in G. Lodolo, 'La tipologia femminile nella commedia elegiaca del sec. XII', in *L'eredità classica* (n. 20 above), 81–100.

comments on the woman's innate weakness which leads her inevitably to adultery (79–80): 'Nature fragili tenuis scintillula cause / sufficit ad culpe prodigialis opus.'[45] The author surely knew Marbod's *Liber decem capitulorum* (c. III: 'De muliere mala'), Hildebert's elegy against women, or Petrus Pictor's *De muliere mala*, where satirical commonplaces, filtered through the Fathers of the Church, were conveniently assembled.[46] However, in the *Milo*, we find an interesting and persistent contrast between the husband, who has to work hard away from his house, and the two lovers, who take advantage of this (67–72). Such solidarity with the cuckold has no parallel, so far as I know, in the works of medieval *clerici*,[47] and is *a fortiori* unique in the spectrum of elegiac comedy. Perhaps some slight sympathy with the poor husbands of certain *matronae* who were fond of famous gladiators, actors, or singers may be traced (perhaps only implicitly) in passages of Juvenal (e.g. 6. 270–85, 359–65), which accordingly may be taken as possible sources of inspiration for our poet.

Concerning point (2), I think it is possible to trace more interesting correspondences with Horace, though, again, not verbal ones.

> tu, cum proiectis insignibus, anulo equestri
> Romanoque habitu, prodis ex iudice Dama
> turpis odoratum caput obscurante lacerna,
> non es quod simulas? metuens induceris atque
> altercante libidinibus tremis ossa pavore.
> quid refert, uri virgis ferroque necari
> auctoratus eas an turpi clausus in arca,
> quo te demisit peccati conscia erilis,
> contractum genibus tangat caput? estne marito
> matronae peccantis in ambo iusta potestas?

You, when you have cast aside your badges, the ring of knighthood, and your Roman dress, and step forth, no longer a judge, but a low Dama, a cape hiding your perfumed head, are you not what you pretend to be? Full of fear, you are let into the house, and you tremble with a terror that

[45] Cf. also 119–22 'Suspectum videt Afra virum nec segnius ardet / et vetitum gaudet amplificare scelus; / in thalamis gaudet regem refovere, laborat / furtis furta, malis accumulare malum'.

[46] See Marbodus (n. 1 above), 97–111; Hildebertus Cenomannensis (n. 22 above), 57–9; Petrus Pictor, *Carmina*, ed. L. van Acker (Turnhout, 1972), 105–16. Among the Fathers of the Church, the most influential upon medieval *clerici* was St Jerome; cf. E. R. Curtius, *Europäische Literatur und lateinisches Mittelalter* (Berne, 1948), 164 n. 7.

[47] In some *fabliaux* the husband triumphs over the adulterers. 'Mais dans les trois quarts des cas les amants triomphent. La majeure partie des histoires ne prennent pas le parti du mari', according to Ménard (n. 37 above), 21.

clashes with your passions. What difference does it make, whether you go off in bondage to be scourged and slashed with the sword or whether, shut up in a shameful chest, in which the maid, conscious of her mistress' sin, has stowed you away, you touch your crouching head with your knees? Has not the husband of the erring matron a just power over both?

This passage of Horace (*Sat.* 2. 7. 53–62) is comparable to the *Milo* not because of the individual scenes here described, but because it displays adulterers of high rank who lose or endanger *de facto* and *de iure* their social position in the attempt to seduce other men's wives, and are thus compelled to hide in strange places and flee from their husbands: an *eques*, or *iudex* (just like the king of the *Milo*) turned into a Dama, a servant,[48] over whom the betrayed husband, if he catches him in the adulterous act, acquires an absolute power (61–2). Yet there is a difference; for in the *Milo* the king, given his position, feels danger not for his life or physical safety, but for his reputation: 'pudor accusat regem; dat terga pudori, / non pene; nec eum vis fugat, immo pudor' (133–4). In this respect another passage of Horace (*Sat.* 1. 2. 127–33) deserves attention:

> nec vereor ne dum futuo vir rure recurrat,
> ianua frangatur, latret canis, undique magno
> pulsa domus strepitu resonet, vepallida lecto
> desiliat mulier, miseram se conscia clamet,
> cruribus haec metuat, doti deprensa, egomet mi,
> discincta tunica fugiendum sit, pede nudo,
> ne nummi pereant aut puga aut denique fama.

No fears have I while screwing her that her husband may rush back from the country, the door burst open, the dog bark, the house ring with the din and clatter of his knocking; that the woman, white as a sheet, leap away, the maid in league with her cry out in terror, she fearing for her limbs, her guilty mistress for her dowry, and I for myself. Clothes dishevelled and bare of foot, I must run off, dreading disaster in purse, person or repute.

The scene in which Milo, coming back from the country, knocks at his own door and then breaks armed into the house, causing the king to flee bare-footed, has so many details in common with this passage, that another possible source, Vitalis' *Geta* (463 ff.) must be considered of lesser importance.[49] The weight of the comparison, however, lies not so much in the fact that the king's

[48] Cf. also Juvenal 6. 44 (an adulterer hides in a chest in order to escape from death).
[49] However, the *Geta* was well-known to Matthew, as a comparison between *Geta* 463 ff. and *Milo* 145–8 clearly shows.

sandals, which in the *Milo* replace the ring (or the sceptre) of the oriental versions, have a parallel in Horace's *pede nudo*,[50] as in the emphasis on the shame which affects the king, humiliating him below the rank of Milo, which may well be a rhetorical development of Horace's final mention of *fama*.

Seen from this perspective—and here I turn to point (3)—the final scene of the comedy (221–42) shows a *iudex*, the king himself, who, for fear of becoming a Dama (that is, in order not to be humiliated by the shame which the revelation of his guilt would bring about) is forced to acquit Milo and give up Afra. Here the limits of his power become perfectly clear, and the author deliberately insists on this point (227–34):

> Quis leo, quis cultor et que sit vinea, sentit
> fossoris solitus rex adhibere vices;
> in commune suum tractari crimen et anceps
> quid tegat et faleret allegoria videt;
> lucescit caligo rei, quod obumbrat amictus
> misterii sensu mens famulante capit.
> Turpe ducem causatur opus culpeque propheta
> coniecturalis arguit ore rubor.

Who the lion is, who the farmer, and which the vineyard, was understood by the king, who was used to play the digger's role; he sees his own crime published abroad and perceives what that ambiguous allegory, covered with ornament, conceals; the obscurity of the situation is unveiled, and his mind, with the aid of reason, grasps what the mantle of mystery overshadows. The dirty deed betrays the duke [i.e. the king] and the blush on his face, proclaiming his suspected guilt, accuses him.

Thus one cannot but ask whether behind Horace something more substantial may be seen, prompting Matthew to modify his main

[50] This element is also found in Apuleius, *Metamorphoses* 9. 20–1, together with other particulars in common with our scene: 'Veneri militabant nudi milites . . . inprovisus maritus adsistit suae domus ianuam . . . iam saxo fores verberat et ipsa tarditate magis magisque suspectus dira comminatur . . . supplicia . . . repandit fores . . . procedit cubiculo, videt sub lectulo soleas incognitas'. Still, a cluster of such details might well be found in any story of adultery then current. There is no need, therefore, to postulate for Matthew a direct knowledge of the *Metamorphoses*; nor am I convinced that Apuleius was read by Petrus Pictor because of the similarities which Hauréau noted between an episode in his *De muliere mala* and *Metamorphoses* 10. 2–12 (although this plot must certainly have circulated until the twelfth century). Cf. B. Hauréau, *Notices et extraits de quelques manuscrits latins de la Bibl. Nat.*, v (Paris, 1893), 220; M. Manitius, *Geschichte der lateinischen Literatur des Mittelalters*, iii (Munich, 1931), 880; van Acker (n. 46 above), xlvii. The detail of the sandals, apparently the most impressive of all, is to be found also in the Spanish version of *Sindbad*, the *Libro de los engaños* (n. 43 above), translated from the Arabic in the thirteenth century.

source so profoundly. I think it probable that the question of how judicial power might be entrusted to a secular authority guilty of the same crimes it was obliged to judge, was a subject of real interest in the twelfth century, as the author's overt denunciation seems to demonstrate (223–6):

> Nil metuit de se qui iudicat, immo reatus
> auctori titulus pro racione sedet;
> regum iustitie preiudicat ausa potestas,
> ius premit et gladius iudicialis hebet.

The judge feels no fear for himself; to the author of the crime, indeed, his position is a substitute for good reason; what kings allow themselves to do endangers their judgment, oppresses law, and enfeebles the sword of justice.

It would be idle to ask whether Matthew of Vendôme (who at different times lived at Tours, Orléans, and Paris) was, when speaking of the king, alluding to Louis VII of France, to whose direct dominions Paris and Orléans belonged, or Henry II, King of England and Duke of Normandy, heir of the county of Anjou and therefore since 1151 lord of Vendôme and Tours. Yet it may not be a mere *lapsus calami* that the king of the *Milo* (*rex* is the obvious translation of *basileus*, the ruler in Constantinople) is twice called *dux* (ll. 72 and 233). What really matters, in my opinion, is that the growing 'bourgeois' groups in the towns of this period, with or without the king's consent, were trying to circumscribe the right of *bannum* of the feudal lords by transferring the administration of justice to local magistrates (the *maior* and the *scabini*), that is, to representatives of the urban community; and this comedy was written in the north of the Loire valley, where precisely at this time these fundamental changes were taking place.[51] If this is so, the medieval reception of classical satire, even in its most artificial form (as in the elegiac comedies), could sometimes affect or reflect the social reality of the day more than we can perceive eight centuries later.[52]

Though I am well aware that many (or most) of the comparisons so far examined are highly speculative, I hope none the less to

[51] See J.-F. Lemarignier, *La France médiévale: Institutions et sociétés* (Paris, 1970), 114–21, 183–4; A. Chédeville, 'De la cité à la ville', in *Histoire de la France urbaine*, ed. G. Duby, ii (Paris, 1980), 175 ff.; and G. Duby and R. Mandrou, *Histoire de la civilisation française*, i (Paris, 1968), 93.

[52] Some comments of Bischoff (n. 21 above, 93 = 269) may be relevant here: 'I think that the close study of the Roman satirists must have helped people to observe the everyday life of their time and encourage . . . realistic tendencies' in contemporary literature.

have proved that Horace and Juvenal furnished the twelfth-century *clerici* not so much with models of language and style, as with a lens through which the shifting forms of historical reality could be brought into focus and interpreted.

7 The 'Pagan Beyond' of Albertino Mussato

MICHELE FEO

1. The Vision: Rejection and Attraction

Though 'enlightened' prejudices against the Middle Ages have at last been overcome, it has not yet been found possible to dismiss the notion that humanism and the Renaissance constituted an advance in the history of European culture. So much so that we find ourselves disconcerted when certain facts or events are not in keeping with this general view. If I had wanted to involve myself in theoretical discussion I could have given my paper a provocative sub-title, such as 'Examples of Reactionary Humanism'. This not being my intention, I shall do my best to keep to the facts and attempt to explain them.

If there exists a typically medieval literary genre—one which was not of course exclusively medieval but which was particularly admired and practised in the Middle Ages—it is that of visions of the afterlife.[1] In order to justify our clichés we tend to characterize

For friendly suggestions and assistance I am indebted to Rossella Bianchi, Elena Giannarelli, Mario Labate, Robert Learmonth, Michael D. Reeve, and Silvia Rizzo. I am equally in the debt of a group of Florentine *dottorandi*, with whom I have discussed the lines of my research in an agreeable meeting: Stefano Brufan, Anna Maria Giannella, Gabriella La Placa, Paolo Maggioni, Antonio Placanica, Luca Robertini, Francesco Stella. Thanks are due to R. A. Scorza for the English translation.

[1] I record here only the most important recent studies from the immense literature on medieval visions and their typology, sociology, and psychology: H. R. Patch, *The Other World according to Descriptions in Medieval Literature* (Smith College Studies in Modern Languages, NS 1; Cambridge, Mass., 1950; repr. New York, 1970); P. Dinzelbacher, 'Die Visionen des Mittelalters. Ein geschichtlicher Umriss', *Zeitschrift für Religions- und Geistesgeschichte*, 30 (1978), 116–28; id. *Vision und Visionsliteratur in Mittelalter* (Stuttgart, 1981); id. 'Körperliche und seelische Vorbedingungen religiöser Träume und Visionen', in *I sogni nel Medioevo: Seminario Internazionale, Roma, 2–4 ottobre 1983* (Rome, 1985), 57–86; C. Carozzi, 'La Géographie de l'au-delà et sa signification pendant le Haut Moyen-Âge', in *Popoli e paesi nella cultura altomedievale, 23–29 aprile 1981* (Settimane di studio del Centro italiano di studi sull'Alto Medioevo, 29; Spoleto, 1983), ii. 423–81; J. Amat, *Songes et visions: L'au-delà dans la littérature latine tardive* (Paris, 1985); K. L. Lynch, *The High Medieval Dream*

visionaries as men of dark medieval religious fervour. By contrast humanists, the lay founders of modern culture, are seen as totally uninterested in visions of the afterlife. Albertino Mussato,[2] however, immediately proves us wrong. For Petrarch Mussato was like a master. He was a great historian, of Livian inspiration; a poet responsible for the revival of tragedy, a genre whose absence from western culture had spanned a millennium; and the first to write a defence of poetry. Nevertheless Mussato—if not the father, then the grandfather of humanism—himself paid homage to the genre of visions of the afterlife, in the form of a short poem of 314 hexameters with the lengthy title *Somnium in egritudine apud Florentiam et commendatio venerabilis patris domini Antonii de Urso episcopi Florentini, cuius beneficio liberatus fuit.*[3]

I must confess that, though I have often convinced myself that I have grasped Mussato's intentions in this work, my conviction has just as often melted away. I once asked an expert in the visions of late antiquity and the Middle Ages whether she had any idea as to Mussato's motives for composing such a work: she answered that she would rather have been surprised had he not done so, since no great man of this era would have willingly excluded a 'vision' from his *curriculum vitae*. This acute observation provides food for thought. Nevertheless one must acknowledge that up until a certain point in his life, Mussato had proudly denied that he would ever indulge in a work of this sort. In a verse letter, which is unfortunately undated but which certainly predates the *Somnium*, he asked a certain Benedetto, a Dominican friar,

Vision: *Poetry, Philosophy and Literary Form* (Stanford, 1988). Cf. also the review by M. Ciccuto, 'Visioni', *Italianistica*, 12 (1983), 345–50. Of great use is the anthology *Visioni dell'aldilà in Occidente*: *Fonti, modelli, testi*, ed. M. P. Ciccarese (Florence, 1987).

[2] On Mussato's background and career, see principally A. Zardo, *Albertino Mussato* (Padua, 1984); M. Minoia, *Della vita e delle opere di Albertino Mussato* (Rome, 1884); M. Dazzi, *Il Mussato preumanista* (*1261–1329*) (Vicenza, 1964); Guido Billanovich, 'Il preumanesimo padovano', in *Storia della cultura veneta*, ii: *Il Trecento* (Vicenza, 1976), 19–110. The whole of *Italia medioevale e umanistica* 28 (1985) is devoted to Mussato and his cultural milieu.

[3] When this paper was given at the Warburg Institute colloquium, I had intended to add to the published version a critical edition of the text. Since an excellent edition has appeared meanwhile I am released from this obligation. See M. Pastore Stocchi, 'Il "Somnium" di Albertino Mussato', in *Studi in onore di Vittorio Zaccaria in occasione del settantesimo compleanno* (Milan, 1987), 41–63. Previous editions, not critical, are those contained in Mussato's works published at Venice, Pinelli, 1636 (sect. vii), pp. 83–90 (preceded by a printing with a variant title-page, of 1635, the unique copy of which is in the British Library: 9150. i. 1), and at Leiden in 1722 in the *Thesaurus antiquitatum et historiarum Italiae* of J. G. Graeve and P. Burmann (vi. 2, App., cols. 63–70), which is an inferior repetition of the Pinelli edition. An Italian translation is given by Dazzi (n. 2 above), 159–68.

whether he would deal with the sighting of a comet and its possible consequences. In his reply, since lost, the friar must have proposed that they should collaborate on the project, either sharing the work equally or dividing it in some other way. Mussato, however, drew back and flatly refused to do so. Imitating Persius' choliambics, he said that he had yet to taste poetry's 'celestial waters' and added, as a Christian reader of the Psalms, that he lacked the wings to fly:[4]

Non ego me sursum tollo nimiosque volatus
experior, casu ne graviore ruam.
Me monet occiduus patrio pro munere quondam
Pheton, et Eridani ripa cruenta vadi.
Sum memor Icarie magno sub teste ruine,
nulla velim pro me nomina dentur aque.
Nec subeo terras, ut opaci scruter Averni
intima, iurate stagna vel atra Stigis.
Non nimis infernos delector visere manes,
unde citus non sic posse redire putem.
Digna Iovis proles nec sum Tyrinthius ille,
mactaret vigilem nec mea clava canem,
nec velut Eneas ulla comitante Sybilla
tutus ab Elysiis credo redire locis.
Stat nobis semper facilis descensus ad ima,
inde pedem tamen est posse referre labor.
Infera Threicius placavit numina vates,
perdidit Eurydicem nec minus ille suam.
Sat contentus ero, media si perfruar aura,
et modo communi cum grege mixtus eam.

I am not one to raise myself up on wings that try to fly too high, only to fall more heavily. I take warning from Phaethon of old, who fell because he took on his father's function, and from the bloody banks of the River Po. I have very strongly in mind also the fall of Icarus, and I want no body of water to be named after me. Nor am I one to go beneath the earth to investigate the dark depths of Avernus or the black, slow waters of Styx, by which the Gods swear. Nor do I think it a great delight to visit the infernal shades, because I do not believe I could soon make good my escape from them. I am not Hercules, worthy son of Jove, to destroy the guardian hound with my club, nor do I think that any Sibyl will bring me safe back as she did Aeneas. Descent to the depths is always easy for us; it is getting back again that is so laborious. Orpheus the poet charmed the Gods below, but lost his Eurydice all the same. I shall be content

[4] *Epist.* 9. 21–40; ed. Graeve-Burmann, cols. 45–6 (the punctuation and spelling are mine).

enough if I enjoy to the full the middle air and travel as one of the common herd.

Upon close examination we find that this refusal is out of phase with the astrological theme proposed. Rather Mussato is using flight as a means of passing from Heaven to Hell so as to launch a polemic against visions of the Underworld. One is certainly aware of a whiff of anti-Dantism. The irony is that Mussato would eventually assume the wings of a dove and do exactly what he here so disdainfully denies. He would rise to the highest Heavens, descend to the depths of Hell, and then see the Elysian Fields. This remarkable contradiction shows that Mussato was apparently not worried about changing his standpoint without revising his writings, in contrast to Petrarch thirty years later.

We can date Mussato's vision precisely. He says that the sun was just over half-way in Virgo, moving towards Libra. Since Virgo runs from 23 August to 22 September, lasting thirty-one days, the sixteenth day brings us to 7 September. Later in the text, however, Mussato also specifies that the vision occurred on the day after the Feast of another Virgin, Maria, in 1319. The Virgin's birthday is 8 September. We have reached 9 September 1319, when Mussato was travelling to Bologna, Florence, and Siena on a diplomatic mission, to seek military aid for his native Padua, which Cangrande della Scala had besieged on 5 August.[5] Around lunchtime he found himself close to Florence and stopped for refreshment at the first tavern he came to. Welcomed by the gracious hostess he entered and sat down. He ate some fresh grapes on an empty stomach and was struck down with a fever, so that he felt cold all over. He was put to bed and covered with a blanket. Hot bricks were applied in an attempt to rid him of his chill, but all was in vain. By the evening his chill had turned into a burning temperature and he feverishly tossed and turned all night. He regained consciousness on the morning of 10 September and asked servants to call for a doctor, who upon his arrival bled him and administered a remedy of almonds and ground spelt. Mussato then sent a servant to the city to alert his friend, Bishop Antonio degli Orsi,[6] who sent a retinue of officials with a

[5] The journey is also recorded in *De gestis Italicorum post Henricum VII Cesarem*, 9. 6: see A. Mussato, *Sette libri inediti del De gestis Italicorum post Henricum VII*, ed. L. Padrin (Monumenti storici, ser. 3, 3; Venice, 1903), 26.

[6] Antonio d'Orso di Biliotto dell'Orso (d. 1320/1) had been bishop of Fiesole until 1309. Dino Compagni describes him as 'di vile nazione, animoso in parte guelfa, e nel vulgo

litter. That evening the party took to the road, travelling at night by the light of the moon (the *Calendario perpetuo* confirms that the moon was then waxing), and they reached Florence early on the morning of 11 September. Mussato was carried to the Bishop's palace and placed on a high bed, more dead than alive. Two doctors ministered to him, the famous Dino del Garbo[7] and another identified only as Lapo. The Bishop, greatly concerned, stayed with him. They made him drink a potion of violets imported from abroad, undressed him, and massaged his body with ointments. At this point Mussato interrupts his account to affirm the veracity of his story, and his metamorphosis begins. The ailing poet's head spins and his sight is dimmed. Feathers sprout from his chest, his arms change into wings, and his legs unite to assume the shape of a tail; his mouth is transformed into a beak. Mussato has become a bird—a dove to be precise—and since *avis* and *columba* are both feminine nouns, he rather coquettishly speaks of himself in the feminine gender from this moment on.

In the shape of a dove Mussato flies out of the palace through an arrow-slit (*rima*) and begins his journey to the world beyond. He first heads for the highest heavens to arrive 'ad medios celi terreque meatus' (line 98), the intermediary zone between Heaven and Earth. Here he sees the spirits of the demi-gods and of those unblemished by sin. Then, driven by a desire to see the bowels of the earth, he flies to the Ocean's edge. In a landscape of sparks and smoke, crowds gather nervously in large numbers to descend into Hell, just like flocks of small birds. The dove circles the area until it finds the entrance to this wretched place and descends by way of a broad, steep path infested with winged serpents. Finally it reaches a large double gate guarded by Cerberus, the terrifying three-headed monster, entwined with serpents and worms. The dove stealthily passes him, but is noticed by the monster, who screams that it will never see the stars

del popolo, ma non di santa vita'; his appointment to the bishopric of Florence is said to have been acquired 'per simonia' (D. Compagni, *Cronica*, iii. 22; ed. I. del Lungo (Rerum Italicarum Scriptores, 2nd edn., ix. 2; Città di Castello, 1907–16), p. 217). Cf. del Lungo's note 11 ad loc.; and F. Ughelli, *Italia Sacra*, iii (2nd edn., Venice 1718), cols. 134–44, no. 43; K. Eubel, *Hierarchia catholica medii aevi*, i (2nd edn. Münster, 1913), 248, 250; R. Davidsohn, *Geschichte von Florenz*, iii (Berlin, 1912), 367, 615–16, and *passim* = Ital. trans., iii (Florence, 1960), 506, 845–7.

[7] A Florentine, he was professor of medicine at Bologna and Siena, and was the author of many works; he died in 1327. Cf. *Nouvelle biographie générale*, xix (Paris, 1857; repr. Copenhagen, 1966), cols. 448–9; M. E. Cosenza, *Biographical and Bibliographical Dictionary of the Italian Humanists and of the World of Classical Scholarship in Italy, 1300–1800*, ii (Boston, 1962), 1547.

again: not even the Queen of Erebus was allowed to return; only two, Hercules and Aeneas, were granted this privilege, and both were of divine birth. The dove ignores Cerberus' threats, continues its flight, and reaches a slimy river to which the spirits are flocking. The dreaded Charon arrives with his boat and shouts at them, telling those who have not been buried to keep away. The dove crosses the river and the swamp, despite being threatened by Charon with his oar, and reaches a dark world lit only by the lightning of an incessant storm. It sees the city of Dis, segmented according to its various punishments,[8] and identifies some of the categories of sinners: the adulterous and the incestuous; the murderers; those who had betrayed their fatherland; the corrupt; and the tyrants and their accomplices. The adulterers and murderers are afflicted with punishments which accord figuratively with their sins (the principle of *contrappasso*). Besides these there are seven more sectors, among which Mussato identifies the parricides and suicides. Further below, at the centre of the Earth, is Orcus, from which there rises a stench which permeates the various precincts. Nearby are Cocytus, Phlegethon, and the Stygian marsh.

The dove flies up and rises above the towers to enter the city of Dis. It sees the King of the Underworld and his wife, Proserpina, horrific sights, and then encounters Rhadamanthus and his companion cross-examining the spirits. Those cleansed of their earthly sins are allowed to cross the blessed threshold of Elysium. A glittering infernal band bearing arms and led by Astaroth—though it is unclear whether they are angels or devils—hold the keys and summon those worthy of leaving the Underworld. Just as the gates are opened the dove shoots like an arrow out of the ivory portal. It sees the stars high in the heavens and the other pole. The sun is about to rise. The dove looks down and sees the flourishing green fields and the happy dwellers in Elysium.

Following the sun's path Mussato reaches the vault of the sky. From there he sees the zones of the earth: Rome and the Capitol; the Italian cities, once the kingdom of Saturn, and Florence, the city which now leaves all others in the shade. Finally he descends towards the Bishop's palace and enters through the narrow arrow-slit. Then having shed his feathers Mussato re-enters his body. The night is over and the early morning sun is shining. Mussato awakes to see the Bishop, the doctors, and the dignitaries

[8] In the marginal notes to his translation, Dazzi (n. 2 above), 163, mentions 'nove case della città di Dite', though Mussato gives no figure.

round his bed. His choice of vocabulary is revealing: 'I awake', *excitor* (260).

The dove's journey into the afterlife was a vision within Mussato's dream of the night of 11–12 September. The new dawn scattered the spirits of the night, and the meaning of the word *excitor* accounts—at last—for the title of Mussato's work, the *Somnium*.[9] Mussato joyfully turns to the Bishop and reassures him that his health has been restored. At this stage the story is over but the poem continues for another forty-four verses, dividing itself into three parts: (1) Mussato invokes the Holy Trinity to accept the poem; (2) he thanks the Bishop for saving his life and wishes him longevity, and then digresses on the Fiesolan origins of Florence, giving rise to a hymn in praise of the beauty, the riches, and the fortunes of this city which reigns over Etruria;[10] and finally (3) he pays political homage to Jacopino, the Vicar of Robert of Anjou, and to his father.[11]

It is immediately apparent that the episodes which frame the vision, namely the journey and the illness which are recounted at the start of the work, and the praises of Florence, her Bishop and her Vicar at the end, are given almost as much weight as the vision itself. There is nothing unusual in a frame having a value and an importance equal to that of its picture, but there must be a reason for such a disproportion of parts.

The *Somnium* was written quickly, and almost certainly in Florence at the time of Mussato's diplomatic mission.[12] Rough-

[9] G. Martellotti, 'Mussato, Albertino', in *Enciclopedia Dantesca*, iii (Rome, 1971), 1067 (and later in his *Dante e Boccaccio e altri scrittori dall'Umanesimo al Romanticismo*, ed. V. Branca and S. Rizzo (Florence, 1983), 455), notes somewhat absently: 'Che si tratti di un sogno è detto nel titolo'.

[10] In l. 288 Florence is called *Parvula Villula*, which Dazzi (n. 2 above), 167, translates as 'il piccolo paesello'. I believe, however, that Mussato is here recalling the legendary name of the ancient nucleus of the city: the 'villa sarnina' or (better) Villa Sarnina, mentioned by Boccaccio (*Comedia delle ninfe fiorentine* (*Ameto*), 38. 56 and 60; ed. A. E. Quaglio (Florence, 1963), 164), on which Mars then conferred the title of Florence after Venus' flowers (ibid. 169); cf. also *Somn.* 299 'consona . . . suo dicta est Florentia flori'. The adjective *sarnino* is, in its turn, related to another tradition: that of the false classicization of the name Arno as Sarno (G. Scalia, '"Arnus"-"Sarnus": Dante, Boccaccio e un abbaglio orosiano', *Studi Medievali*, ser. 3, 20 (1979), 625–55).

[11] On the Brescian Jacopino di Ponte Carali, see R. Davidsohn, *Forschungen zur Geschichte von Florenz*, iv (Berlin, 1908), 547; id. (n. 6 above), iii. 629 n. 4 = Ital. trans. iii. 864–5 n. 4. His father Fiorino had been *podestà* of Florence in 1309, at the time when Mussato had held the office of executor of the *Ordinamenti di Giustizia* (ibid. iii. 335–6, 370–2 = Ital. trans. iii. 466–7, 510–3).

[12] Pastore Stocchi (n. 3 above), 42, rightly maintains that it could not post-date 1320, since Bishop Antonio, for whom the work augurs a long life, died at the end of 1320 or early in 1321.

nesses and illogicalities of expression[13] betray a certain haste in its composition.[14] Furthermore there are many literal quotations taken from classical poets,[15] as well as metrical anomalies[16] and repetitions.[17]

If we examine the vision's typology and phenomenology, we find that the *Somnium*'s papers are 'all in order'. The vision is closely linked to a dream and even occurs within a dream.[18] Illness is normally a prerequisite for a vision,[19] and Mussato's dream is preceded by a mortal illness, his body in a grave state of weakness. Visions should manifest themselves within a 'sacred place',[20] normally a church, but as a layman Mussato substitutes the Bishop's palace for a church. They are also a means of rewarding holy men for their many days and nights spent in prayer.[21] Mussato was not exactly an ascetic, but he clearly wanted to retain this association with the concept of prayer and so

[13] Two examples: (1) In the initial chronological indication (1–4) 'Iam Phebi currus exceperat hospita Virgo / vixque sue mediam partem peragraverat edis, / acelerans breviare dies sub lege perenni / et sibi contigue venienti cedere Libre') the subject is always *Virgo*; yet if according to the Ptolemaic system it is correct to say that the Virgin receives the chariot of the Sun into her 'house', it is the Sun rather than she who traverses this 'house'. (2) On waking after the first night of fever, Mussato cries to his servants that he feels a fire burning within and that his tongue is blackened (43 'exusto nigrescit lingua palato'); unless he had looked at himself in a mirror, however, he could not have been aware of the latter detail (nor is Pinelli's rendition of the phrase as an interrogative a solution).

[14] Pastore Stocchi (n. 3 above), 44–51, has demonstrated the existence of two phases in the work's composition. Of these the more mature form (the vulgate) has not, however, eliminated all the original defects, nor could it have done so, except at the cost of a fundamental rewriting.

[15] On this see below. Proof of his 'abilità di mosaicista' (Dazzi (n. 2 above), 106) are the 386 verses of the Ovidian cento written in 1318 (ed. Graeve-Burmann, cols. 69–76).

[16] They are indicated by Pastore Stocchi (n. 3 above), 52–3. To them should be added a further two, problematic instances: 201 'Quis penas referat? Quis tantum *exprīmere* possit' of the entire MS tradition (M. D. Reeve has brought to my notice the possibility of emendation to *exprōmere*); and 231 'Portant centuplices vario *fŏramine* claves', which contrasts with the corrected 95 'audeoque antique resecare *fŏramina* rime' (but the reading *varioque* of MS Columbinus 7-5-5 in Seville, not recorded in the apparatus—provided it is not a conjecture by the copyist—re-establishes the norm, even if, in turn, it creates a difficult relation between the attribute *centuplices* and the ablative of quality *vario foramine*).

[17] Some examples: *hospita* occurs three times (1, 13, 21) in the fifth foot; in the fifth also *sidera* is repeated in quick succession (37 and 39). Various verse endings are reiterated: *meatus* is used twice (10, 98) as is *ydri* (171, 214), *ministri* three times (81, 282, 309), and likewise *sinistra/is* (211, 222, 230); the neighbouring phrases *reverentior oris* (287) and *reverentibus ore* (297) are, if not identical, very similar. The phenomenon is also seen at verse beginnings: *Virgo* (6, 132), *excitor* (34, 260), *parvula* (129, 288). The anaphoric reiterations, on the other hand, have an expressive function: *aspicio* (92 and 96, though not at 237); *scire velim* (105) and *scire volo* (108); *gens* (293–4), and obviously the anadiplosis linking 300 and 301, 'oscula *dextre*; / *dextre*, inquam . . .'.

[18] Cf. Dinzelbacher, 'Körperliche' (n. 1 above), 58–60; it is significant that the title of Amat's work is *Songes et visions* (n. 1).

[19] Dinzelbacher, ibid. 70–5. [20] Ibid. 81–2. [21] Ibid. 82–4.

incorporated two invocations into the *Somnium*: the first to the Virgin when he was frightened by Cerberus (131–2); the second, as we have seen, to the Trinity (271–6).

Let us now turn to the poet's metamorphosis into the shape of a dove (86–91):

> In caput evolvor suppressaque lumina condo;
> tuncque meum video subito plumescere pectus
> brachiaque extensas sese convertere in alas
> astrictosque pedes unam coniungere caudam;
> os quoque mutatum rostro se extendit acuto.
> Iam sum avis et facto gaudens nova tegmina quasso.

My head spins and my sight is dimmed; I see my breast growing feathers, my arms becoming extended wings, and my feet joining together in a tail; while my mouth is transformed and elongated into a sharp beak. I have become a bird and, delighting in the fact, I flutter my new plumage.

Strangely enough this passage recalls the description of the witch Pamphile in Apuleius' *Golden Ass* (3. 21),[22] though Mussato could not have known this text, if the only extant manuscript in the early Trecento (Laurentianus LXVIII 2, 11th c.) was indeed not brought from Montecassino to Florence until many years later.[23] It is more plausible that Horace provided the model (*Odes* 2. 20. 9–12):

> iam iam residunt cruribus asperae
> pelles, et album mutor in alitem
> superne, nascunturque leves
> per digitos umerosque plumae.

Now rough skin grows on my legs; while above I am transformed into a white bird, and light plumage grows on my arms and shoulders.

This probability is due not so much to precise verbal dependence,[24] but rather to a similarity in the symbolic content of the texts. Horace's flight 'per liquidum aethera' (2) represents poetry's triumph over time and death (8, 'nec Stygia cohibebor unda').

[22] Martellotti, *Dante* (n. 9 above), 455–6.

[23] R. Sabbadini, *Le scoperte dei codici latini e greci ne'secoli XIV e XV* (Florence, 1905–14; new edn., E. Garin, ii, Florence, 1967), 202; E. C[asamassima], in *VI Centenario d. morte di G. Boccaccio: Mostra di documenti e edizioni*, i (Certaldo, 1975), 129–31, no. 105; P. K. M[arshall], in *Texts and Transmission: A Survey of the Latin Classics* (Oxford, 1983), 15–16.

[24] Analogies may be recognized in Mussato's use of Horace's anaphoric *iam*; in the flight of the two birds over cities and landscapes; and in Horace's final exhortation to banish funeral laments, which is comparable with the 'cohibete dolorem' of *Somn.* 269. Horace claims immortality (6–7 'non . . . obibo'), and Mussato declares himself alive (270 'aspiro'). The comparison was suggested to me by A. Placanica.

Another more literal quotation recalls the so-called *Somnii elegia* attributed to Ovid (*Amores* 3. 5). Here we find exactly the same expression, 'rostro . . . acuto' (39), as in line 90 of the *Somnium*. The pseudo-Ovidian *elegia* narrates an erotic dream. Mussato's borrowing, therefore, either illustrates his open-mindedness towards indiscriminately appropriating quotations from visionary writings or, possibly, represents a sly way of deceiving the reader. The verses which follow, however, immediately and forcefully draw us back to sources which are more in keeping with the spiritual nature of Mussato's vision (94, 96):

> . . . flave species michi visa columbe . . .
> aspicio celos, celum super ardua tendo.

I seemed, in the shape of a tawny dove . . . to contemplate the heavens, to travel the lofty places above the sky.

These words recall David's wish to rise to the heavens: 'Quid dabit mihi pennas sicut columbae et volabo . . .?' ('Oh that I had the wings of a dove! for then I would fly away', Psalm 54. 6), a wish which has its fulfilment in an oracular passage: 'Si dormiatis inter medios cleros, pennae columbae deargentatae, et posteriora dorsi eius in pallore auri' ('Though ye have lain among the *cleros*, yet shall ye be as the wings of a dove covered with silver, and her feathers with yellow gold', Psalm 67. 14) and ends in certainty with the words: 'ascendisti in altum' ('Thou hast ascended on high', 19). However we interpret *cleros*,[25] the Psalms allude to the notion of the transformation, through a dream, of the human body into the shape of a dove—the golden-yellow of its tail corresponding with the colour *flavus* in Mussato[26]—and its flight to the heavens. If this allusion to the scriptures, unique as far as I can see, was really intended, the implication is that Mussato wanted to allude to the arcane relationship which he believed to exist between pagan and biblical poetry.

The dove's flight has been in fact the flight of the spirit, as is evident from this bird's traditional symbolic association with the

[25] The interpretation preferred by the Church Fathers is 'terminos', i.e. the two Testaments. Cf. Augustine, *Enarrationes in Psalmos*, ed. E. Dekkers and I. Fraipont (Corpus Christianorum, ser. lat. xxxix; Turnhout, 1956), 879–85; Jerome, *Opera*, i. 1, (ibid., ser. lat. lxxii; Turnhout, 1959), 214; Cassiodorus, *Expositio Psalmorum*, ed. M. Adriaen (ibid., ser. lat. xcvii; Turnhout, 1958), 591; for all the Fathers, R. Bellarmino, *Explanatio in Psalmos* (2nd edn., Paris, 1633), 470–1. The possibility should not, perhaps, be entirely excluded that Mussato, given that he was lodged in a prelate's house, may have wished to understand *cleros* as 'priests'.

[26] For the meaning 'yellow' of *pallor*, see M. Feo, '"Pallida no, ma più che neve bianca"', *Giornale storico della letteratura italiana*, 152 (1975), 321–61.

concepts of simplicity and virtue.[27] This is made clearer by Mussato's return at the conclusion of his flight, when he relinquished a second metamorphosis, and in keeping with a traditional schema made the spirit re-enter the body in the form of a dove.[28] The metamorphosis was therefore an illusion, a form of ecstasy and a departure of the spirit from the body, which in its absence was motionless as though dead.

In my opinion this relationship between the vision and illness is a figure of human powerlessness when confronted with such phenomena. One could even call it the flight from rationality. Weakened by physical illness, the intellectual faculties lose their autonomy and are enslaved to the process of revelation. They can only witness, endure, and record it. Visions provide access to a truth which is not gained through reason, but through grace and divine authority. This is why humanism, even as it denies nothing of Christian truth, does not identify its model hero as a visionary but in the pragmatic, energetic, and tireless *exemplum* of Hercules.[29] The vision must by its very nature contain a moral message or a far-reaching admonition, whether minatory or hopeful. In this respect the *Somnium* is disconcerting. *Video*, in all its analogues and synonyms, is the word recurring most often—more than thirty times in 150 lines. Mussato's vision is one of pure contemplation:

[27] Cf. Amat (n. 1 above), 373–4.

[28] Into the theme of the soul's repossession of the body, writers sometimes weave novelistic elements. The monk of Vaucelles begs the accompanying angel not to make him go back (P. G. Schmidt, 'Die Vision von Vaucelles (1195–1196)', *Mittellateinisches Jahrbuch*, 20 (1985), 162, l. 45), whilst Thurkill is made to return in furious haste by Saints Michael and Julian, so as to prevent those who are attempting to reanimate his lifeless body from killing him with imprudent remedies; the author then adds humorously that he does not know 'quo ordine' the re-entry has occurred (*Visio Thurkilli, relatore, ut videtur, Radulpho de Coggeshall*, ed. P. G. Schmidt (Leipzig, 1978), 36 = *Die Vision des Bauern Thurkill*, ed. P. G. Schmidt (Leipzig-Weinheim, 1987), 80). Yet already in previous centuries, other visionaries had given 'realistic' explanations of the phenomenon: in order to receive Fursa's soul (7th century), the subject's breast had split open (Ciccarese (n. 1 above) 222), whilst his contemporary, Barontus, had re-entered less bloodily 'per os' (ibid. 266). As regards Mussato, those who have read 256–8 'Desilio extimplo conspecta presulis aula / et tenui rima positis de corpore plumis / ingredior corpusque meum mortale subinto' literally have understood that the dove re-entered the body via a *rima*. This interpretation, however, involves clumsy hypotheses about which *rima* of the body is intended. It is preferable to suppose that Mussato is silent on the problem and by *rima* means that of the palace through which the dove had flown away.

[29] On the marked presence of the myth of this hero in Florentine civilization, it is sufficient to recall Coluccio Salutati's *De laboribus Herculis* (ed. B. L. Ullman (Zurich, 1951)); see also principally L. D. Ettlinger, 'Hercules Florentinus', *Mitteilungen des Kunsthistorischen Institutes in Florenz*, 16 (1972), 119–42; M. M. Donato, 'Gli eroi romani tra storia ed "exemplum": I primi cicli umanistici di Uomini Famosi', in *Memoria dell'antico nell'arte italiana*, ed. S. Settis, ii (Turin, 1985), 129.

it neither encourages nor discourages the reader. His descriptions of crime and punishment, subjects of potentially dark and terrifying rhetoric (if otherwise used in the *Ecerinis*!), would not even affect the most innocent or sensitive of children.

2. *Back to Virgil*

I believe that an aid to understanding the *Somnium* is possibly to be found in a passage in Mussato's *Epistle* 18 on the defence of poetry. This was written in 1316 and therefore predated the vision.[30] Upholding the variety of poetry Mussato wrote (60–64):

> Nunc tibi quo metuas fert horrida Musa timores,
> nunc lenis placidis mulcet tua pectora verbis;
> ethica nunc, nunc phisis erit; nunc vera methesis
> Cociti nigramque Stigem iuratque videtque,
> surgit ad empyreum nunc velocissima celum.[31]

It is not easy to interpret lines 62–3. According to the excellent editor *vera methesis* means 'true divination' and *Cociti* is indeclinable but used here in the accusative ('she swears and sees Cocytus and the black Styx'). He cites another example of *Cociti* as an indeclinable noun in line 199 of the *Somnium*: 'Sunt prope Cocciti, Phlegeton Stigieque paludes'. I remain unconvinced for two reasons: (1) at *Somnium* 199 *Cocciti* can be understood perfectly well as the nominative plural, that is, as a standard declined form

[30] I accept the dating of Dazzi (n. 2 above), 191.

[31] E. Cecchini, 'Le epistole metriche del Mussato sulla poesia', in *Tradizione classica e letteratura umanistica*: *Per Alessandro Perosa*, i (Rome, 1985), 111–12; see also comment on p. 101. Petrarch too, in contrasting the polychromy and variety of Homer and Virgil with the monotone of David, distinguished the description of otherworld phenomena as a specific field, perhaps a canonical genre, of poetry: see *Bucolicum carmen*, 1. 78–88:

> necnon et triplicis sortitos numina regni
> expingunt totidem varia sed imagine fratres;
> sceptiferum summumque Iovem facieque serena,
> inde tridentiferum moderatoremque profundi
> ceruleumque comas medium, fuscumque minorem;
> torva latus servat coniunx, aterque paludis
> navita Tartaree piceas redit itque per undas,
> tergeminusque canis latrat; tum dura severis
> pensa trahunt manibus fixa sub lege sorores;
> quin etiam Stygias eterna nocte tenebras
> anguicomasque simul Furias . . .

It is well known that Petrarch's ideas about poetry depend directly on Mussato's (see Giuseppe Billanovich, 'Pietro Piccolo da Monteforte tra il Petrarca e il Boccaccio', in *Medioevo e Rinascimento*: *Studi in onore di Bruno Nardi*, i (Florence, 1955), 1–76, esp. 18–20).

like *Stigie paludes* instead of *Stigia palus*. This example therefore does not confirm that *Cociti* in line 63 is indeclinable;[32] (2) if *Cociti* were in the accusative Mussato would be saying that the gods swore by Cocytus and the Styx, whereas according to Virgil (*Aeneid* 6. 324), Servius, Seneca (*Hercules furens* 711–13), and Mussato himself, the gods swore only by the Styx. I need only recall the expressions 'iurata deis Stix' (*Epistle* 18. 15), 'quid superos iurare Stigem miraris?' (ibid. 65), and finally 'iurate stagna vel atra Stigis' (*Epistle* 9. 28). In fact *Cociti* is in the genitive and is dependent on *methesis*. The passage means:

Now frightening, the Muse summons the terrors you feared; now she gently caresses your mind with soothing words; now she is Ethics, now Physics, and now the True Knowledge of Cocytus, and sees and swears by the black Styx; now she flies rapidly to the heaven of the empyrean.

The adjective *vera* reminds us how it is stated twice in the *Somnium* (85, 275) that the episodes recounted represent not only truth, but sacred truth. It would thus appear that Mussato's poetic vision of the Underworld should not be classed with ethical works,[33] but that it belongs to a doctrinal cognitive sphere.

It is here that the *Somnium* begins to reveal its secrets. It aims to provide a *vera mathesis*, a true understanding of the world beyond as opposed to some other misguided form of *mathesis*. Mussato writes as the heir to a rich tradition of medieval visions which he uses eclectically, selecting and modifying his sources. His relationship with the tradition is variegated and only apparently extravagant. The elements he chooses are carefully and consistently selected. Though his work is firmly rooted in the medieval genre, he seeks to restore it to its classical origins. Some examples will serve to clarify his aims and methods.

It is significant that Mussato retains the *Ichform*, using the first

[32] For Pastore Stocchi (n. 3 above), 53, the *Cocciti* of l. 199 is a genitive dependent on *paludes*—a solution less forced than that of the indeclinable form. The translation of Dazzi (n. 2 above), 192, is, however, incongruous: 'ora sarà Etica, ora Fisica, ora vera Scienza; ora giura per il nero Stige formato dal Cocito . . .'

[33] In the medieval *accessus*, poetry 'ethice supponitur'. The most important studies of this subject are *Accessus Ovidiani*, ed. G. Przychocki (Rozprawy Akademii Umiejętnósci w Krakowie, 49; Cracow, 1911); *Accessus ad auctores*; Bernard d'Utrecht; Conrad d'Hirsau, *Dialogus super auctores*, ed. R. B. C. Huygens (Leiden, 1970); E. A. Quain, 'The medieval *Accessus ad auctores*', *Traditio* 3 (1945), 215–64; F. Ghisalberti, 'Mediaeval Biographies of Ovid', *Journal of the Warburg and Courtauld Institutes*, 9 (1946), 10–59; B. Nardi, 'Osservazioni sul medievale *accessus ad auctores* in rapporto all'Epistola a Cangrande', in *Studi e problemi di critica testuale: Convegno di studi di filologia italiana nel Centenario della Commissione per i testi di lingua (7–9 aprile 1960)* (Bologna, 1961), 273–305.

person in the narrative,[34] but rejects a prose format. Medieval visions recounted in prose had come to be accepted as part of the epistolary genre. For example Vaucelles' vision of 1195/6 is in the form of a letter, as is confirmed by his regular use of the *cursus*. Rather than adopt the rhythmical cadences of epistolary prose, Mussato preferred the classical dignity of the hexameter. In so doing he followed in the wake of another great visionary work, the *Visio Wettini* of Walahfrid Strabo, to whom Mussato was also indebted for the acknowledgements with which he ends his poem. Technically the *Somnium* is therefore a verse letter and as such can legitimately be regarded as a humanist work. Horace and Ovid, the ancient masters of the epistolary genre, had left their heirs free to choose between hexameters and elegiac couplets along with a vast range of subject matter. Mussato's choice of verse is therefore of fundamental importance, since it differentiates his work from those of prose authors, who were required to maintain a degree of historical truth. It allows him to tap instead the more complex 'poetic truth', of which he was the first humanist to make a reputation as a great theorist and apologist.[35] Finally it is worth pointing out that Mussato's vision was not part of an epic poem. This may have been chance, but if it was, it was perfectly in keeping with his own reservations about the function of epic within fourteenth-century society.[36]

If we consider the structure of Mussato's journey, we see that it is very similar to that in the vision of the young Englishman Orm (1126).[37] The purpose of this comparison is not to suggest a direct relationship between these texts, but rather to illustrate how they share certain features peculiar to the genre. Orm loses consciousness in the night, and his body becomes rigid as though he is dead. At first his servants vainly attempt to revive him. In the end they are amazed to see him move as though he is awakening from a deep sleep. Orm recounts that he was carried on high by the Archangel Michael 'above the stars' ('super sydera'), where

[34] I borrow the term from German writers on visions, for example P. G. Schmidt, 'Vaucelles' (n. 28 above), 157; and H. Knittel, ed. and tr., Walahfrid Strabo, *Visio Wettini*: *Die Vision Wettins* (Sigmaringen, 1986), 23.

[35] Cf. A. Galletti, 'La "ragione poetica" di Albertino Mussato e i poeti-teologi', in *Scritti varii di erudizione e di critica in onore di Rodolfo Renier* (Turin 1912), 331–59; G. Vinay, 'Studi sul Mussato, i: Il Mussato e l'estetica medievale', *Giornale storico della letteratura italiana*, 126 (1949), 113–59; Dazzi (n. 2 above), 108–23; G. Ronconi, *Le origini delle dispute umanistiche sulla poesia* (*Mussato e Petrarca*) (Rome, 1976).

[36] I deal with this question in a work now in the course of publication: 'Il poema epico latino nell'Italia medioevale'.

[37] H. Farmer, 'The Vision of Orm', *Analecta Bollandiana*, 75 (1957), 72–82.

he saw God the Father, Christ, the Angels, the Virgin Mary, and the Twelve Apostles. He then descended beneath the world, into a very deep and dark valley full of stinking smoke and darkness ('subtus mundum, ad vallem profundissimam ac tenebrosissimam, plenam fetenti fumo et caligine'), where he saw Satan and the spirits subjected to punishments and torments. From here, he went ever upwards from the north to the midday region of the sun ('ab aquilone semper ascendendo ad meridianam plagam solis'), and proceeded 'ad paradisum', the delightful abode of great throngs of *sancti*, the blessed dead. From Paradise he rose again to the Heavens and was readmitted to the presence of God the Father. Finally, on his return to England via Jerusalem, he saw Rome from on high.

We should note that both versions share this phenomenology of trance followed by reawakening, along with further novelistic elements. Moreover it is remarkable that the passage of Mussato's dove, comprising five stages, is identical to that of Orm's spirit: (1) on high towards the Heavens; (2) down to the depths of Hell; (3) to a southern abode of the blessed, namely Elysium; (4) again to the vault of the Heavens; (5) and finally home by way of Rome. But even more remarkable is the fact that apart from this skeletal framework there is no similarity between the pious young Orm's vision and that of Mussato. Mussato makes no reference to God or to the celestial court, nor any edifying comments. He paganizes the other world, wishing to see it through the eyes of the ancient poets. Since he does not know Greek his principal, though not totally exclusive model, can only be Virgil.

Mussato particularly admired Virgil, whom he placed at the apex of poetic achievement. He had once lent his copy of Virgil to Guizzardo, a friend from Bologna who, along with Castellano, had written a commentary on the *Ecerinis*.[38] Then, much as today, friends can hardly be trusted when it comes to returning borrowed books, and this was the case with Mussato's Virgil. He therefore wrote Guizzardo a request in verse, adding:[39]

> Virgilius, thalamo mecum versatus in uno,
> tempore quo Patava pulsus ab urbe fui

[38] *Comentum super tragedia Ecerinide editum a magistro Guiçardo Bononiensi trivialium doctore et Castellano Bassianensi artis gramatice professore ab aliisque artistis examinatum et probatum*, in A. Mussato, *Ecerinide tragedia*, ed. L. Padrin, with a study by G. Carducci (Bologna, 1900), 67–247.

[39] *Epist.* 14. 5–10, ed. Graeve-Burmann, col. 50 (the spelling and punctuation are mine). The epistle is dated 1318, and translated into Italian by Dazzi (n. 2 above), 174.

exul ad externas ultro se contulit oras,
exilii penas sustinuisse volens.
In patriam redii, redeat securus et ipse:
et comes et civis sit velut ante fuit.

When I was expelled from the city of Padua, Virgil, as companion of my bedchamber, betook himself voluntarily to exile on other shores because he wished to comfort me in the pains of banishment. Now that I am back in my native city, let him also return to me safe and sound and be my companion and fellow citizen as once he was.

I do not believe that anyone has so happily extended the topos of the dialogue between reader and book into the ambiguous territory of the bedchamber.[40] This 'declaration of love', both touching and humorous, should have been enough alone to merit an entry for Mussato in the *Enciclopedia Virgiliana*.[41]

Fleeting glimpses of Virgil are found in many medieval visions,[42] no less than in the infernal scenes described in epics: for example Joseph of Exeter, Walter of Châtillon, and Enrico da Pisa.[43] But Virgil's presence in the *Somnium* is particularly marked. There are many expressions and *iuncturae* derived from the ancient poet, and in this respect Virgil was a veritable mine into which Mussato freely delved. Mussato's borrowings cover the whole rhetorical range. At the most fundamental level there are formulaic borrowings like 'Ditis regna' (106; cf. *Aeneid* 6. 269 'Ditis . . . regna'); 'oppressere . . . *crudeli funere*' (191; cf. *Eclogue* 5. 20

[40] C. Bec, 'De Pétrarque a Machiavel: A propos d'un "topos" humaniste (le dialogue lecteur/livre)', *Rinascimento*, ser. 2, 16 (1976), 3–17; see also my note in *Annali della Scuola Normale Superiore di Pisa*, Cl. Lett. e filos., ser. 3, 7 (1977), 1697–8.

[41] Mussato's name is also absent from D. Comparetti, *Virgilio nel Medioevo* (Florence, 1872; new edn., G. Pasquali, Florence, 1937–41; 4th edn., ibid. 1981).

[42] E. Norden, in P. Vergilius Maro, *Aeneis Buch VI* (Leipzig, 1903; 8th edn., Darmstadt, 1984), 7, was too hasty in denying Virgil's influence on medieval apocalyptic writings, as has been shown by P. Courcelle, 'Les Pères de l'Église devant les Enfers virgiliens', *Archives d'histoire doctrinale et littéraire du Moyen Age*, 22 (1955), 5–74. Courcelle concludes (65): 'Une enquête plus étendue et plus minutieuse montre que les auteurs de *Visions* chrétiennes, plusieurs hagiographes grecs et latins et quantité de poètes ont eu recours, pour la représentation de l'au-delà, à l'imagerie virgilienne. L'apocryphe *Vision de saint Paul*, qui date au plus tard du IVe siècle, a fondé une tradition où la topographie virgilienne des Enfers, l'énumération des supplices, le cliché des cent bouches, la description des Champs Élysées sont aisément reconnaissables. Même des Pères de grand renom n'ont pas hésité, surtout s'ils étaient doués de sens poétique, à admettre pour leur compte l'existence des fleuves infernaux et à concevoir le Paradis sous l'aspect des bois fortunés que parcourut le héros troyen'. Cf. also Ciccarese (n. 1 above), 56–7 and *passim*.

[43] For Joseph and his *Ylias* (ed. L. Gompf (Leiden-Cologne, 1970)) see *Annali della Scuola Normale Superiore di Pisa*, Cl. Lett. e filos., Ser. 3, 9 (1979), 1837–8; for Walter of Châtillon see *Alexandreis* (ed. M. L. Colker (Padua, 1978)), 10. 6–125; for Enrico see *Liber Maiolichinus*, 2937–3039 (ed. C. Calisse (Rome, 1904), 111–14).

'exstinctum . . . *crudeli funere*'; 'satus Anchisa' (141; cf. *Aeneid* 6. 331 'Anchisa satus'; 'remigio alarum' (123; cf. *Aeneid* 1. 301 'remigio alarum' and 6. 19 'remigium alarum', both in identical places in their respective lines). This last image had already been 'spiritualized' by the Church Fathers in their description of the wings of Christian Charity.[44] Even Virgil's *tegmen fagi* (*Eclogue* 1.1), under which Tityrus was stretched, finds its way into Mussato's vision of Hell: it was enough for Mussato to paint it black, which is there a more appropriate colour (167 'sub tegmine nigri').[45]

At first sight Mussato's '*per* cuncta *silentia* noctis' (45) appears to be a more exalted allusion (cf. *Aeneid* 2. 255 '*per* amica *silentia* lunae', and Statius, *Thebaid* 2. 58 '*per* . . . *silentia* lunae') aimed at creating a comparable atmosphere of suspense. However, the interlacing with the often abused phrase *silentia noctis* (cf. Lucretius, Ovid, Valerius Flaccus, Paulinus of Nola, Paulinus of Périgueux, Sedulius, Rabanus Maurus, Walahfrid Strabo, Ermoldus Nigellus, Abbo of Saint-Germain, Henry of Auxerre, and Ligurinus),[46] tells us that we are a long way from the varied or concealed forms of imitation which would subsequently form the basis of Petrarch's theory and practice.[47] Virgil's famous image of the 'linguae centum' is not found in Mussato's *Somnium*:[48]

> Non, mihi si *linguae* centum sint oraque centum,
> *ferrea* vox, omnis scelerum comprendere formas,
> omnia poenarum percurrere nomina possim. (*Aeneid* 6. 625–7)

If I had a hundred mouths, a hundred tongues and a voice of iron, I could not include every sort of crime, or list the name of every retribution.

> Non ego cuncta meis amplecti versibus opto,
> non, mihi si *linguae* centum sint oraque centum,
> *ferrea* vox. (*Georgics* 2. 42–4)

44 Courcelle (n. 42 above), 10–11. Cf. *Vita s. Guthlaci*, 33 (Ciccarese (n. 1 above), 386).

45 Though it should be said that the combination *tegmine/a* + a spondaic adjective or noun as hexameter-clausula is very common after Virgil and throughout late antique and medieval Latinity: O. Schumann, *Lateinisches Hexameter-Lexikon*: *Dichterisches Formelgut von Ennius bis zum Archipoeta*, (*MGH Hilfsmittel*, iv, 1–6; Munich, 1979–83), v, 362–3.

46 Ibid. v. 155–6.

47 Cf. G. Martellotti, '"Similitudo non identitas": Alcune varianti petrarchesche', *Rivista di cultura classica e medievale*, 19 (1977) (= *Miscellanea di studi in memoria di Marino Barchiesi*), 491–503 (now in *Scritti petrarcheschi*, ed. M. Feo and S. Rizzo (Padua, 1983), 501–16); M. Feo, 'Petrarca, Francesco', in *Enciclopedia Virgiliana*, iv (Rome, 1988), 64–6.

48 On its long history see P. Courcelle, 'Histoire du cliché virgilien des cent bouches (*Georg*. ii, 42–44 = *Aen*. vi, 625–7)', *Revue des études latines*, 33 (1955), 231–40; see also Corippus, *Iohannis* 1.22h–23 (ed. M. A. Vinchesi (Naples, 1983), 41, 88–9.

I cannot hope to cover in verse all the wide expanse, not even if I had a hundred tongues, a hundred mouths and a voice of iron!

Nevertheless Mussato's admission of the powerlessness of the human tongue (178–9) must bear some relationship to this precedent:

> Quotque alie tenebris pene versantur in illis,
> cor nullum caperet nec *carnea lingua* referret

How many other punishments dwell in that darkness, no heart can compass, no human tongue describe,

as shown by the substitution of *carnea* for Virgil's *ferrea*.[49]

Occasionally the obviousness of Mussato's borrowings is intended to be an allusion of a more or less explicit kind—or rather, in Mussato's verses there is both a semantic and an emotive echo of his Virgilian model. See, for example, lines 74–5: '*totum*que a vertice *corpus* / ad pedis extremum *dudum*que volubile *lustrant*' (cf. *Aeneid* 8. 153 'iam*dudum* et to*tum lustrabat* lumine *corpus*'). Here Mussato's description of the doctors' anxiety as they examine his inert corpse shares in the intensity of Evander's feelings when he sees Aeneas' features and is reminded of Anchises. And how can one not be aware that in his vote of thanks to Jacopino, Mussato's 'Si bene quid merui' (313) is almost a verbatim quotation of the lovelorn Dido's words to Aeneas, 'si bene quid de te merui' (*Aeneid* 4. 317), and retains something of their Virgilian weight? At a significant point in Mussato's poem—his journey from Hell to Elysium—I believe that the literal borrowings are intended to draw the reader's attention to the extent to which he adheres to, yet at the same time departs from, the structure of his model: '*portis* divellor *eburnis*' (236; cf. *Aeneid* 6. 898 '*porta*que emittit *eburna*'); and '*camposque* virentes / . . . video' (241–2; *cf. Aeneid* 6. 677–8 '*camposque* . . . nitentis / . . . ostentat'); but the whole account of the Christian Elysium (237–47) recalls Virgilian vocabulary (sun, stars, meadows, fields, delight, costumes, light, choirs, grass, and games).

As part of this process of breaking down and recomposing his model, of dialogue and imitation, Mussato even brought himself to the extreme of Christianizing 'strips' of the lacerated Virgil's flesh, as in line 272 where Christ is described as the 'certa Dei

[49] In a prose epistle written for carnival probably in the late Trecento, I have found the expression 'que nunquam posset exprimere *lingua carnis*' (M. Feo, 'Il nemico e l'alleato', in *Manipulus florum*: *A Eugenio Garin* (Pisa, 1980), 2, ll. 25–6).

proles celo demissa supremo' ('the true offspring of God sent from on high'). The first hemistich is virtually identical with Virgil's 'deum certissima proles' (*Aeneid* 6. 322) in reference to Aeneas. The second hemistich reproduces the famous line from the fourth Eclogue, 'iam nova progenies caelo demittitur alto', which an entire tradition of medieval literature had made its own.[50] Mussato's intention in using Virgil's prophetic words was not to correct him—that would almost have been an act of impiety—but to make him speak the truth which was within him without his knowing it.

The geography of Mussato's Hell (City of Dis, rivers, marsh, Elysium) also comes from Virgil. He calls the souls *manes*, exactly as a pagan would.[51] Cerberus, Charon, Pluto, Proserpina, and Rhadamanthus are all pagan deities. Thus the *Somnium* made its modest contribution to the assimilation of classical antiquity, something which has preoccupied generations of artists and writers. Mussato, however, was not a slavish imitator and at times departed from Virgil either for dogmatic reasons or to pursue other sources, or even to allow his imagination free rein. We shall see how he revised, or rather enriched Virgil by means of Cicero. For reasons of orthodoxy he placed the suicides among the sinners (192–3), whereas in Virgil (*Aeneid* 6, 434–6) they are innocent spirits ('insontes').[52] Moreover he introduced Astaroth, transforming a Jewish goddess into a male demon.[53] He even changed

[50] Cf. K. Strecker, 'Jam nova progenies caelo demittitur alto', *Studi medievali*, NS 5 (1932) (= *Virgilio nel medioevo*), 167–86.

[51] But *anime* in l. 222. The souls of the dead are also called both *manes* and *anime* in the *Liber Maiolichinus* 2976, 2977, 2988, 3007. On the adoption of the term *manes* into Christian culture see C. Pascal, *Le credenze d'oltretomba nelle opere letterarie dell'antichità classica*, i (Catania, 1912; 2nd edn., ibid., 1923), 127–41.

[52] On this Mussato is in accordance with Christian doctrine: cf. Courcelle (n. 42 above), 19. For the correction to Virgil made by Petrarch in *Africa* 6. 11–15 see note ad loc. in N. Festa's edition (Florence, 1926), 133; see also E. Leube, 'Petrarca und die alten Götter: Zum Bild der antiken Mythologie in der *Africa* und im übrigen lateinischen Werk des Dichters', *Romanistisches Jahrbuch*, 11 (1960), 93; Feo, 'Petrarca' (n. 47 above), 63b, where the reference to Walter of Châtillon should be deleted.

[53] *Astroth* at the end of the hexameter *metri causa* instead of *Astaroth*. The demon, 'regi fidissimus', is described as 'hirsutus galea, curvum sublimior arpen / attollens dextra, peltam retinente sinistra', at the head of a company bearing arms of shining white (227–30). The biblical Astaroth may be identified with the Phoenician goddess Astarte. Cf. G. L[evi] D[alla] V[ida], in *Enciclopedia italiana*, v (Rome, 1930), 72–3; F. Nötscher and Th. Klauser, in *Reallexikon für Antike und Christentum*, i (Stuttgart, 1950), cols. 806–10. For attestations amongst Christian writers see H. Plenkers in *Thesaurus linguae latinae*, ii (Leipzig, 1902), col. 945. In the *Passio s. Bartholomaei* Astaroth is an Indian idol in which a demon lodges (*Acta Apostolorum apocrypha*, ed. R. A. Lipsius and M. Bonnet, ii. 1 (Leipzig, 1898 = Hildesheim, 1959), 128–33); the story is also recalled by Johann von Neumarkt in a letter to Petrarch (*Petrarcas Briefwechsel mit deutschen Zeitgenossen*, ed. P. Piur and K. Burdach

the location of the Elysian Fields. In Virgil they are situated within the confines of Hell, with the exit at its border. In Mussato, however, the gates are located between Hades and Elysium—a name which Mussato does not in any case use. They lie in the southern hemisphere under the second pole star, and have their own sky (235–8):

> Egredior Partha satis ocior ipsa sagitta
> et gressum extorquens portis divellor eburnis.
> Aspicio celi specimen stellasque micantes
> decernoque polum tali regione secundum.

I shoot faster than a Parthian arrow and tear a way through the gates of ivory. I discern the heavens and the shining stars and the second pole in that region.

At first sight it is not clear whether Mussato has tampered with Virgil or misinterpreted him. Virgil (*Aeneid* 6. 641) and subsequently Claudian in the *De raptu Proserpinae* (2. 282–4) located the Elysian Fields under the Earth, with their own underground sky and their own sun. Mussato, however, was probably attempting to rationalize this concept of the other sky by transferring the Elysian Fields to the antipodes over the earth's crust. This controversy actually dated back to late antiquity and had preoccupied both Servius (*Aeneid* 6. 640) and Macrobius (*Commentarii in Somnium Scipionis* 1. 9. 8). Servius ingenuously attempted to resolve the problem by offering several alternative readings of Virgil, suggesting that the Elysian Fields were either in the underworld, or in the Fortunate Isles, or in the sphere of the moon ('campi Elysii aut apud inferos sunt aut in insulis fortunatis aut in lunari circulo'); Macrobius, as we shall see, sought to compromise between poetic imagery and philosophical meaning.[54] Mussato had therefore to deal not with a pure but with an already interpreted Virgil. Furthermore, during the course of the centuries the pagan Elysium and the Christian terrestrial paradise had almost become assimilated: the one as a figment of the poetic imagination, originally located in Hell and the western hemisphere; the other located at a remote setting in the east, or indeed in the antipodes.[55]

(Berlin, 1933), 137). In the Italian Renaissance, as is well known, Astarotte had a resounding success as the philosopher-devil in Pulci's *Morgante* (21. 49; 25. 119–66, 232–44, 277–332; 26. 79–89; 27. 51; 28. 33).

[54] Cf. Courcelle (n. 42 above), 34–5.

[55] On the location of the Elysian Fields, the myth of the earthly paradise, and the various attempts to find it, see A. Graf, 'Il mito del Paradiso terrestre', in his *Miti, leggende e*

Mussato also borrowed from other classical authors. From Seneca's vision of Hell[56] he derived the description of Charon's matted beard (153 'barba impexa'; cf. *Hercules furens* 766—Mussato's derivation was not in fact from Virgil, *Georgics* 3. 366, as proposed by the annotations in the Pinelli edition); from Statius' *Thebaid* (12. 816) came Mussato's 'vive precor' (303). Cerberus' 'whimper' (131 '*vagitum dedit ille* canis') is a bizarre literal borrowing from Ovid (*Heroides* 11. 85 '*vagitus dedit ille* miser'), where the unfortunate is a child. If, as I believe, Mussato's memory was prompted by Virgil's 'vagitus . . . ingens' (*Aeneid* 6. 426), which also concerns children but which immediately precedes the description of Cerberus, one must admit that his compositional technique was a little coarse. On the other hand one should give credit to Mussato's memory and his familiarity with the classical literary heritage, especially since he was writing while away from his library. Formulae like 'mundi machina' (132) or 'est locus' (195) had by then become commonplaces,[57] but others were not so obvious. The clausula 'Iovis altitonantis' (140) is particularly perplexing, given that it comes from a fragment of Ennius (*Annales* 541 V.) referred to in Varro's *De lingua latina* (7. 7), a work which lay undiscovered in Montecassino in 1319.[58]

superstizioni del medio evo, i (Turin, 1892), ix–238 (2nd edn., Turin, 1925, 1–175); R. R. Grimm, *Paradisus coelestis, Paradisus terrestris: Zur Auslegungsgeschichte des Paradies im Abendland bis um 1200* (Munich, 1977); G. Gatto, 'Le voyage au paradis: La christianisation des traditions folkloriques au Moyen Age', *Annales: Economies, Sociétés, Civilisation*, 34 (1979), 929–42; Carozzi (n. 1 above), *passim*; G. Orlandi, 'Temi e correnti nelle leggende di viaggio dell'Occidente alto-medievale', in *Popoli e paesi* (n. 1 above), ii. 523–71; M. Degli Innocenti, 'Ancora sulla letteratura dei viaggi oltremondani: La "Leggenda del Paradiso terrestre"', *Italia medioevale e umanistica*, 29 (1986), 63–88 (with previous bibliography). Amat (n. 1 above), 398–401, has shown how the syncretism Paradise-Elysian Fields-Garden of Eden was favoured by poetic clichés of Virgilian origin, describing Elysian or bucolic landscapes.

[56] Attention has been directed to this by Pastore Stocchi (n. 3 above), 42.

[57] For the first see Schumann (n. 45 above), iii. 267–8 ('machina mundi') and 453 ('mundi machina'). For the second, ibid. ii. 190–1: 31 instances are recorded, to which others could easily be added, e.g. Ps.-Cyprian, *Carmen de resurrectione* 194; id. *De Pascha* 1; Lactantius, *De phoenice* 1; Walter of Châtillon, *Alexandreis* (n. 43 above), 10. 58; Goffredo da Viterbo, *Pantheon* part 1, in Graf (n. 55 above), 2nd ed., 151; Alanus of Lille, *Anticlandianus* 1.55 (ed. R. Bossuat, Paris, 1955); H. Favoli, in W. Ludwig, *Litterae neolatinae* (Munich, 1989), 156; cf. also H. Walther, *Initia carminum ac versuum medii aevi posterioris latinorum* (2nd edn., Göttingen, 1969), nos. 5720–8; id. *Proverbia sententiaeque latinitatis medii aevi*, i (Göttingen, 1963), nos. 7576–9; L. Bertalot, *Initia humanistica latina*, i (Tübingen, 1985), nos. 1550–1. The vernacular equivalent of the syntagma occurs in Dante as 'luogo è' (*Inf*. 18. 1, 20. 67, 34. 127; *Purg*. 7. 28). It should also be noted how frequently this expression introduces the descriptions of Inferno or the earthly Paradise.

[58] The 11th-cent. MS (Laurentianus LI.10), father of all surviving codices, was brought to Florence by Boccaccio in 1355. Cf. Sabbadini (n. 23 above), i. 30–1; E. C[asamassima] (n. 23), i. 136–8, no. 110; L. D. R[eynolds], in *Texts* (n. 23), 430–1.

In conclusion, despite substantial differences, one can say that Virgil's pagan Hell was the major point of reference for Mussato, and primarily its description in the sixth book of the *Aeneid*—that great sixth book which has been so much admired and discussed during the course of the centuries[59]—and almost certainly that of the fourth book of the *Georgics* and perhaps the *Culex*. Mussato's *vera mathesis* is both Virgilian and Christian. The devout and barbaric *mathesis* of medieval ascetics and mystics had simply to be corrected by pre-Christian wisdom.

3. Classicism versus Dante

If we now consider chronology, we see what everybody has seen, that the *Somnium* is contemporary with Dante's *Comedia*. No reader of the *Somnium* has felt comfortable with this knowledge, and a satisfactory solution to the problem has yet to be found. I feel justified in using the word 'comfortable', since it is embarrassing to compare one of the greatest of literary creations with indifferent occasional verses. But compare them we must, since it seems that this is what Mussato would have wanted.

There are certain objective parallels in the lives and aims of both Mussato and Dante.[60] As contemporaries they exalted the cultural achievements of the Italian *comuni* to opposite heights. Dante, an impoverished noble, a Ghibelline uprooted from his city, looks out to the world; Mussato, a commoner who rose to the ranks of the nobility and a Guelph, remains tenaciously rooted in his city. Dante clearly understands the potential of the vernacular and 'creates' Italian literature: Mussato stays within the Latin

[59] Cf. once more Courcelle (n. 42 above), and now also his *Lecteurs païens et lecteurs chrétiens de l'Eneide*, i: *Les Témoignages littéraires* (Institut de France, Mémoires de l'Academie des inscriptions et belles lettres, NS 4; Paris, 1984). For Mussato, *Aeneid* 6 has an importance that goes beyond its being a model of a transcendental vision. From 6. 853, for example, he derives the epiphonemic formulation of the mission of the Roman People, transferring it to the Emperor Henry VII: 'Parcere subiectis et debellare superbos' (*Epist.* 5. 78; ed. Graeve-Burmann, col. 42). On this see *Dizionario degli autori italiani* (Messina-Florence, 1973; 2nd edn., ibid. 1974), 992. An extensive series of correspondences between the *Somnium* and *Aeneid* 6, which, however, is not always above criticism, is given by Dazzi (n. 2 above), 72–8.

[60] M. Feo, 'Tradizione latina', in *Letteratura italiana* (ed. A. Asor Rosa), v; *Le Questioni* (Turin, 1986), 314; and see also M. Pastore Stocchi's insightful essay 'Dante, Mussato e la tragedìa', in *Dante e la cultura veneta: Atti del Convegno di Studi* (*Venezia, Padua, Verona, 30 marzo–5 aprile 1966*) (Florence, 1966), 251–62. Among previous studies of the problem of the relations between Dante and Mussato the following may be cited: M. Dazzi, 'La fama del Mussato', *Rivista d'Italia*, 19. 1 (1916), 107–11; A. Belloni, 'Dante e Albertino Mussato', *Giornale storico della letteratura italiana*, 67 (1916), 209–64; M. Dazzi, 'Il Mussato storico', *Archivio Veneto*, ser. 5, 6 (1929), 452–4.

tradition and plants the seeds of humanism. Mussato, the author of the *Ecerinis*, senses the world as a tragedy: Dante, through the *Comedia*, constructs his metaphysical optimism. In 1315 Mussato was awarded the poet's garland in his native city: by contrast Dante died in Ravenna, his wish to receive the garland—'il cappello'—in his 'bel San Giovanni' unfilfilled. Generations of scholars since the unification of Italy have dreamed of finding a point of connection between these parallels, but no work of either contains any explicit reference to the other. We do not know if they ever met, but it is unthinkable that they were not at least aware of one another. Mussato was in Florence in 1309 as executor of the *Ordinamenti di Giustizia*, and must have been well-informed about affairs in the city: Dante was in Verona under the patronage of Cangrande della Scala, who was the principal enemy of the Paduans. For once, in 1310, they had something in common when the Emperor Henry VII represented the political aspirations of both these men, but their paths always led in different directions.

One who admired them both, the Bolognese teacher Giovanni del Virgilio, attempted to draw them closer together. In the Spring of 1319 he invited Dante to write a Latin epic, with the promise that it would be read and commented on to the students and that Dante would be crowned poet in Bologna. He was therefore promising Dante exactly what Mussato had already achieved. When Dante declined, Giovanni stated that if he insisted on refusing he would quench his thirst in the 'Phrygian Musone', namely the Paduan stream. He meant allegorically that he would open a verse dialogue with Mussato, as indeed he did after Dante's death.[61] The implication is that Dante was capable of understanding Giovanni's allusion to Mussato. Furthermore it seems that in *Paradiso* 9. 28–32 Dante included a polemic against the *Ecerinis* at the point where Cunizza, the sister of the grim tyrant who was the protagonist of Mussato's tragedy, denies implicitly that her brother was born of the devil. Here it is perfectly legitimate to speculate on what was left unsaid. Silences of this sort between 'giants' who obstinately follow different paths

[61] The most recent editions of the poetic correspondence between Giovanni del Virgilio and Dante are those by G. Albini, repr. G. B. Pighi (Bologna, 1965), which also contains Giovanni's eclogue to Mussato; by E. Cecchini, in D. Alighieri, *Opere minori*, ii (Milan-Naples, 1979), 645–89; and by G. Brugnoli and R. Scarcia (Milan-Naples, 1980). To the flood of Italian articles and polemics on the problems raised by the correspondence should be added K. Krautter's balanced work, *Die Renaissance der Bukolik in der lateinischen Literatur des XIV. Jahrhunderts: von Dante bis Petrarca* (Munich, 1983).

are in themselves full of eloquence, as anyone who has lived through the Cold War knows.

It cannot be mere coincidence that Mussato wrote his work on the afterlife in Florence, where Dante was detested, and in 1319 when the *Inferno* and *Purgatorio* had already been published and the *Paradiso* was awaited. Vladimiro Zabughin saw the *Somnium* as 'a masterpiece of malice and finesse',[62] naturally against Dante. It may be that at times Mussato's writing is subtle, as Zabughin suggests, but on the whole he is rough and feeble. Even to compete with the *Comedia* is going too far. Mussato failed to understand that the *Comedia* is not a medieval stereotype. Dante brilliantly transformed the visionary's status from that of an individual who passively receives the truth to that of one who realizes a cognitive process, whereby he is guided to truth by reason. The vision becomes a function of rationality, almost a substitute for it. Consequently Dante omitted all the standard preamble of illnesses and dreams, features which would tend to weaken the validity of this process of reasoning and consciousness. In Dante's hands the vision becomes the Virgilian or Homeric journey in which the hero or the sage advances 'by ever more exalted stages of experience' towards self-fulfilment and the attainment or rather the conquest of *ἀρετή*, *virtus*.[63] This is why Dante journeyed in the Beyond with Virgil. Mussato, however, as Zabughin wittily observed, went on a 'pleasure trip'.[64] On this trip Mussato, like a good tour guide, carefully drew attention to the various sights and claimed even to correct in a sly way Dante's mistakes. Being blind to the great poetry of the *Comedia*, Mussato examines individual details with an almost philological eye, of a kind that would have pleased Petrarch.

Despite the fact that there exists an artistic chasm between these works, scholars have often seen or wanted to see points of formal similarity between the *Somnium* and the *Comedia*, and especially the *Inferno*.[65] Some of these may be the result of

[62] V. Zabughin, *L'oltretomba classico, medievale, dantesco nel Rinascimento*, i (Florence-Rome, 1922), 34.

[63] On this Stoic basis of the *Aeneid* and its influence upon Dante see E. Fraenkel, 'Carattere della poesia augustea', *Maia*, 1 (1948), 245–64 (and later in his *Kleine Beiträge zur klassischen Philologie* (Rome, 1964), ii, 209–29, esp. 225).

[64] Zabughin (n. 62 above), 34.

[65] Dazzi, 'La fama' (n. 60 above), 107–11; A. Belloni, 'Una visione dell'oltretomba contemporanea alla dantesca', *Rassegna Nazionale*, Ser. 2, 36 (nov.–dic. 1921), 20–31. Other 'conati paradanteschi'—to use Dazzi's aggressive expression—have been sought for in the *Soliloquium* 4 (of 1321) and in the *De lite inter Naturam et Fortunam* (of 1327): cf. A. Moschetti, 'Il "De Lite inter Naturam et Fortunam" e il "Contra casus fortuitos" di

common classical sources, but others in my opinion cannot be dismissed so easily. See for example 'nunquam consueta sub astra redibis' (*Somnium* 136; cf. *Inferno* 3. 85 'non isperate mai veder lo cielo'); 'stat semper resonans imber' (163; cf. *Inferno* 5. 31 'la bufera infernal che mai non resta'). However, the frequency of similar episodes in both the poems must have some significance.

Dazzi discovered a possible criticism of Dante on Mussato's part in lines 195–6 of the *Somnium*.[66] Mussato states that the abode of the traitors at the centre of the earth has always been known as Orcus: Dante, however, called it Antenora after the Trojan hero Antenor (*Inferno* 32. 88), the mythical founder of Padua, who according to Servius (*Aeneid* 1. 242) and Dares Phrygius (*Excidium Troiae* 39–42) had sold his country to the Greeks. The Paduans must have disliked the name Antenora, and Mussato tacitly accuses Dante of committing a historical error. Such 'teasing' on minute details was not rare; Mussato had heavily criticized Tanto, the Chancellor of the Venetian Republic, for a metrical solecism.[67] By the same token Dante is also probably drawing attention to an error in the geography of the *Ecerinis* when, in the speech where Cunizza contests her brother's demonic birth, she states that the Ezzelino's fortress rises on 'un colle . . . non . . . molt'alto' (*Paradiso* 9. 28). Mussato, however, described an 'excelso . . . colle' (*Ecerinis* 8–9). It may be that modern philologists let their imagination run riot, but how can we otherwise explain that exactly the same point was made by another poetic admirer of the della Scala, namely Ferreto de' Ferreti?[68]

When Cerberus warns the dove that only two mortals, Hercules and Aeneas, have ever left the Underworld alive, Mussato's words clearly recall Charon's reproach to Aeneas. As we have seen, Mussato had already alluded to this in his letter to Fra Benedetto. In addition the *Somnium* is probably levelling criticisms at Dante's description of the beginning of his voyage when, aghast at

Albertino Mussato', in *Miscellanea di studi critici in onore di Vincenzo Crescini*, pre-publication extract (Cividale del Friuli, 1913), 10–11; final edition (Cividale del Friuli, 1927), 574–5; E. Cavallari, *La fortuna di Dante nel Trecento* (Florence, 1921), 344–5; Dazzi (n. 2 above), 78–9; Martellotti, *Dante* (n. 9 above), 456.

[66] Dazzi (n. 2 above), 72 n. 80. Both Martellotti, loc. cit. (n. 65) and Pastore Stocchi (n. 3 above), 42, express perplexity on the matter.

[67] Cf. G. Monticolo, 'Poesie latine del principio del secolo XIV nel codice 277 ex Brera al R. Archivio di Stato di Venezia', *Propugnatore* NS 3, 2 (1890), 244–303.

[68] On the whole question see E. Raimondi, 'Dante e il mondo ezzeliniano', in *Dante e la cultura veneta* (n. 60 above), 51–69 (later under the title 'L'aquila e il fuoco di Ezzelino', in his *Metafora e storia: Studi su Dante e Petrarca* (Turin, 1970), 123–46, esp. 138–41).

what he had undertaken, he recalls that such a privilege had only previously been granted to Aeneas and to St Paul: 'Io non Enea, io non Paulo sono' (*Inferno* 2. 32). Paul, for his faith, was transported to the Third Heaven (2 Cor. 12: 2–4), but did not in fact descend into Hell. Therefore modern scholars maintain that Dante was confused about the famous apocryphal *Visio Pauli*.[69] When he went back to the topos of Charon's threats and the poet's fear of the journey, Mussato probably chose to adhere firmly to classical heroes and so consciously omitted St Paul as a way of indirectly criticizing Dante. Nor can we exclude the possibility that Mussato was censuring Dante for making Virgil himself descend into Hell when he was still alive (*Inferno* 9. 16–30; 12. 34–45; 21. 61–3). This 'fantasy' finds no support among ancient authors and as such has intrigued many commentators, beginning with Dante's own son Pietro. He held that the idea was 'non vera' and only justifiable 'tropologice, idest moraliter'.[70]

Mussato's most frequent corrections to the *Comedia* relate to the topography of the Underworld. Dante places the souls' point of embarkation for Purgatory at the mouth of the Tiber (*Purgatorio* 2. 100–5), but fails to specify the location of the entrance to Hell. Mussato, however, proves to be an outstanding expert in this field. He describes a landscape which with its smoke and flames recalls the Phlegraean Fields and is located at the Ocean's edge (105–17):

> Scire velim, qua parte ime sint intima terre.
> Tunc video fumos et Ditis regna superbi
> surgentesque rubras ad sidera summa favillas.
> Scire volo, qua parte animas ad Tartara portant,
> et, si forte vie quavis michi sorte paterent,
> infaustos tentare locos et regna videre.
> Cumque satis cavi, venit innumerabile vulgus,
> qua iacet Occeanus mergens Yperiona fundo;
> inde ruunt populi sine fine, sine ordine, quotquot
> tempora narrares non mille per annua natos.

[69] The first to show awareness of the matter was A. F. Ozanam, *Dante et la philosophie catholique au treizième siècle* (2nd edn., Paris, 1845), 344–5. Cf. A. D'Ancona, *I precursori di Dante* (Florence, 1874), 43–5 (repr. in his *Scritti danteschi* (Florence, 1912), 37–9). The thesis that Dante knew the *Visio Pauli* was later sustained with new arguments against sceptics and detractors by the greatest expert on and editor of that work: T. Silverstein, 'Did Dante know the Vision of St. Paul?', *Harvard Studies and Notes in Philology and Literature*, 19 (1937), 231–47.

[70] Cf. V. Ussani, 'I viaggi di Virgilio nel sottoterra', in *Wirtschaft und Kultur: Festschrift zum 70. Geburtstag von Alfons Dopsch* (Baden b. Wien-Leipzig, 1938), 604–10.

More avium veniunt modicarum, ut forte cicade;
murmur apum sonat ista cohors celeratque meatum;
namque loco se queque timet succedere tarde.

I seek to know in what region are the innermost parts of the lower world. Then I see smoke and the kingdoms of the proud Dis and the rosy sparks rising to the highest stars. I want to know from what region souls are carried to Tartarus and—if the Fates allow the ways to lie open for me—to explore those unhappy places and see those kingdoms. Whilst I look at the scene very cautiously, an innumerable crowd arrives at where Oceanus lies, that sees the sun going down into its depths; thence run people without number, disorderly, so many that you could not count them in a thousand years. They come like little birds, or like cicadas; their masses make a noise like the buzzing of bees and rush along as if afraid of coming late to the spot.

The Hell of the *Visio Pauli* is close to the Ocean,[71] but the sources used by Mussato appear to have been both more dignified and more authentic. In time to come Petrarch's investigations would reach the same conclusions: Servius (*Aeneid* 6. 107) said that Homer, like Virgil, located Ulysses' *νέκυια* in Campania, despite his story that the event took place at the Ocean's edge; furthermore, according to Claudian the entrance to Hell was in Gaul, again close to the Ocean (*In Rufinum* 1. 123–8).[72] One comes away with the impression that Mussato, long before Petrarch, had carried out the same research and reached similar conclusions.

Before descending into Avernus Mussato's dove rose to the heavens where it saw the *manes* of the demi-gods and the innocent, those allowed to enjoy the 'serene' sky and to gather flowers in the meadows (98–101):

Efferor ad medios celi terreque meatus.
Semideos manes video nunquamque nocentes
permissos illic celo gaudere sereno
et roseos placidis flores decerpere campis.

[71] *Visio Pauli* (red. i) 1 'Et duxit eum super flumen magnum aque. Et interrogavit eum Paulus quis esset hic fluvius. Et dixit angelus: "Hic est Oceanus, super quem girant sidera celi, et circuit orbem terre". Et vidit ibi locum terribilem, et non erat lumen in illo, sed tenebre et mesticie et suspiria. Et nomen loci eius Cochiton; et alia tria flumina que confluunt secum Stix, Flegeton et Acheron' (ed. T. Silverstein, *Visio Sancti Pauli*: *The History of the Apocalypse in Latin together with Nine Texts* (Studies and Documents, iv; London, 1935), 153). Cf. id. 'The Vision of Saint Paul: New Links and Patterns in the Western Tradition', *Archives d'histoire doctrinale et littéraire du moyen âge*, 26 (1959), 226–7).

[72] For Petrarch see M. Feo, 'Inquietudini filologiche del Petrarca: il luogo della discesa agli Inferi (Storia di una citazione)', *Italia medioevale e umanistica*, 17 (1974), 115–83.

For many scholars[73] this description represents Limbo. In Dante, however, Limbo is located underground. Doctrinally this is a Thomist creation[74] which incorporates Virgilian images and ideas. In some respects it imitates the Virgilian foyer to Hell, inhabited by the spirits of the innocent; furthermore, it resembles Elysium in its account of the spirits of the great, and the brightness and the sweetness of its landscape. Dazzi, curiously enough, believes that Mussato's supposed Limbo—Mussato does not even use the term—is found 'su alte cime fiorite',[75] being probably misled by the mention of fields covered with flowers. In fact there is no doubt that this passage alludes to a celestial space and not to terrestrial mountain peaks. We would have to ask Mussato himself for further elucidation of the possibility that meadows and flowers lie in that space; but he may have been thinking of something like the Milky Way. There is nothing strange in the concept of two paradises, one terrestrial and the other celestial: this was a commonplace during the Middle Ages. According to Dante (*Monarchia* 3. 16. 7) one is the symbol of this life's happiness that comes from the practice of the virtues, and the other of eternal beatitude. What is original about Mussato is that his paradises are both pagan. If the model for his terrestrial paradise was Virgil's Elysium, its counterpart receives its humanistic authentication from Cicero's *Somnium Scipionis*,[76] in which the heavens are the abode of those who have served their fatherland (3. 5): 'sic habeto: omnibus qui patriam conservarint, adiuverint, auxerint, certum esse in caelo definitum locum; ubi beati aevo sempiterno fruantur.' For his part Mussato places there generically souls of the demi-gods[77] and the innocent.

Macrobius had already found this contradiction between Virgil and Cicero troublesome, and had sought to eliminate it by giving to Virgil's text two senses, one literal and the other symbolic (*Commentarii in Somnium Scipionis* 1. 9. 8–10):

Vergilius . . . licet argumento suo serviens heroas in inferos relegaverit, non tamen eos abducit a caelo, sed *aethera* his deputat *largiorem et nosse eos*

[73] The idea is Belloni's (n. 65 above), 24.

[74] F. Montanari, 'Limbo', in *Enciclopedia Dantesca*, iii (Rome, 1971), 651–4.

[75] Dazzi (n. 2 above), 73.

[76] On the history of this work in a Christian context, see P. Courcelle, 'La Postérité chrétienne du "Songe de Scipion" ', *Revue des études latines*, 36 (1958), pp. 205–34; and now A. Hüttig, *Macrobius im Mittelalter: Ein Beitrag zur Rezeptionsgeschichte der Commentarii in Somnium Scipionis* (Diss. Freiburg, 1989).

[77] It is curious that Zabughin (n. 62 above), 34, should translate 'semideos manes' as 'diavoli'.

solem suum ac sua sidera profitetur, ut geminae doctrinae observatione praestiterit et poeticae figmentum et philosophiae veritatem.

Virgil . . . although he consigns his heroes to the underworld in accordance with his plan, does not remove them from the sky. He gives them 'a broader ether and the knowledge of their own sun and their own stars', showing in this way his command of both the poetic and the philosophic disciplines, imaginative and truthful respectively.

The fact that the spirits of the blessed in Virgil's Elysium continue to exercise their physical faculties, exactly as they had done on earth, was seen by Macrobius as allegorically alluding to the concept that they do not renounce their task of governing human society after they have entered the highest sphere (*sphaera* ἀπλανής).

Twenty years after Mussato, Petrarch, in the first book of his *Africa*, took over the heavenly territory, reserving it exclusively for those who had died for their fatherland. In a dream, Scipio sees the spirit of his father who carries him through the heavens (175–9) to a place where there is 'pura dies, quam lux eterna *serenat*' (219), 'in regione *serena*' (496). There Scipio sees his uncle, Fabius Maximus, Caius Gracchus, Paulus Aemilius, the Kings of Rome (with the exception of Tarquin the Proud, who is in Hell), the three Horatii, and countless others. All were men of action, statesmen and generals. Still later, in 1350, in a verse letter to Virgil (*Familiares* 24. 11) he asked him, among other things, whether he lived in Tartarus, Elysium or in 'celi regio tranquilla' where the stars look benignly on the serene spirits, now that the King of Kings has conquered Tartarus. Here Petrarch is implicitly expressing two of his greatest wishes: (1) that Virgil might be beatified and enter the Christian paradise; and (2) that the great Italic poet might dwell, thanks to his poetry, among the benefactors of their fatherland.[78] Petrarch had gone right back to Cicero, but this notion of 'serenity' also owes something to Mussato. Petrarch ascribed this concept also to Italy (*Epystolae* 2. 11) which at the time—free of pollution—could still be regarded as an earthly paradise. There is much evidence to suggest that Petrarch had carefully read the *Somnium*.[79]

[78] Cf. Feo, 'Petrarca' (n. 47 above), 63 b, although I was then not yet aware of the second important implication.

[79] A long, literal borrowing from *Somn.* 74–5 '*totumque a vertice corpus / ad pedis* extremum dudumque volubile *lustrant*' may be seen in *Africa* 1. 200–1 '*totumque a vertice corpus / lustrat ad* usque *pedes*', where a comparison with the Virgilian model (*Aen.* 8. 153—see above) strongly suggests that Petrarch's direct source was Mussato rather than Virgil. From the

Finally Mussato's greatest disagreement with Dante on the structure of the Underworld lies in his total exclusion of Purgatory. In the *Somnium* the cleansed spirits pass directly from Hades into the Elysian Fields. Purification of the spirits was already in Virgil (*Aeneid* 6. 730–51), but Mussato does not give an account of this process. He does not contest the concept of expiation in the afterlife, but denies the existence of a zone or a *regno* dedicated to this function, an area commonly known from the end of the twelfth century as Purgatory.[80] Not only had Dante made the concept of Purgatory his own, not only had he given it a dimension equal and opposite to that of the infernal precincts, but he had taken the 'heterodox' liberty of making it the setting for his terrestrial paradise, which he located at its summit. Despite eliminating Purgatory from his scheme, however, Mussato did not totally reject the tripartite division of the world beyond. Rather than Hell, Purgatory, and Paradise we have the Heavens, Hell, and Elysium.

Other visionaries also denied the existence of Purgatory, for example Bonvesin de la Riva and Giacomino da Verona. Mussato was therefore not alone. One even finds a tinge of embarrassment among Dante's partisans at the existence of the second part of the *Comedia*. One of the two epitaphs carved on his tomb mentions as Dante's works only the *Monarchia* and the *Comedia*, and the latter is indicated by allusion only to the *Inferno* and *Paradiso*: 'Iura Monarchie, superos, Phlegetonta lacusque' ('The powers of Monarchy, those above, Phlegethon, with the lakes').[81] Moreover in a poem of the 1340s, Moggio Moggi described a public exposition of the *Comedia* given in Verona by Pietro, Dante's son. Moggio's summary is somewhat confused, but his tendency was clearly to classicize and to paganize the poem. It is

terrifying figure of 213–14 derives the description of Persephone in *Africa* 3. 244 as *inamena* and in *Bucolicum carmen* 1. 83 as *torva*, in contrast with the traditional depiction of the maid as beautiful. In *Il Petrarca ad Arquà*: *Atti del Convegno di studi nel VI centenario (1370–1374), Arquà Petrarca, 6–8 Nov. 1970* (Padua, 1975), 123 n. (and cf. *Quaderni petrarcheschi*, 4 (1987), 345 n. 11), I have sought the classical source for Persephone in this infernal guise in Horace, *Odes* 2. 13. 21; it may also be that it might somehow (as A. M. Gianella has suggested) originate from the designation ἐπαινή which is standard in Homer (*Iliad* 9. 457, 569; *Odyssey* 10. 491, 534, 564; 11. 47). See also above, n. 31.

[80] J. Le Goff, *La naissance du Purgatoire* (Paris, 1981; Ital. trans., Turin, 1982).

[81] Cf. A. Campana, 'Epitafi', in *Enciclopedia Dantesca*, ii (Rome, 1970), 711, who believes the author to be Bernardo Scannabecchi. G. P. Marchi, in *Vestigia*: *Studi in onore di Giuseppe Billanovich*, i (Rome, 1984), 417–28, attributes it rather to Rinaldo Cavalchini da Villafranca.

noteworthy that, like Mussato's dove, Moggio flies directly from Hell to Elysium.[82]

Now, if we seek to discover the reasons for these apparent criticisms of Dante by an intellectual who was always *engagé*, unaccustomed to write evasively, we must turn yet again to the chronology and to the cultural and political context of the time. Mussato was travelling towards a friendly Florence in early September: in August Cangrande della Scala had laid Padua under siege. Dante, far from a Florence hostile to him, had recently glorified Cangrande (*Paradiso* 17. 85–7):

> Le sue magnificenze conosciute
> saranno ancora, sì che' suoi nemici
> non ne potran tener le lingue mute.

This could well be a further reference to Mussato, who despite his enmity had as historian exalted Cangrande's deeds. Dante was also prepared to address his Thirteenth Epistle to Cangrande when dedicating the *Paradiso* to him.[83] A few months earlier Giovanni del Virgilio in Bologna, a city equidistant from Verona, Padua, Florence, and Ravenna, had eagerly raised the issue of Dante's coronation, but Dante had declined any compromise: the coronation must be in Florence and nowhere else, on account of the *Comedia* and for no other work. In defiance of Dante, Mussato threw his *Somnium* on to the scales like the sword of Brennus. The *Comedia*, he claimed, was replete with errors. This was enough to illegitimize Dante's aspirations to the poet's laurel, or at least to provide a convenient excuse for the refusal to those in cultural and political authority, short-sighted and hostile as they were. In essence Mussato condemned the *Comedia* just as it would be condemned in the maturity of humanism in the fifteenth century. We need only recall the famous passage from the *Dialogi ad Petrum Histrum*:[84]

> Ut enim a Dante incipiam . . . nonne illum plerumque ita errantem videmus, ut videatur rerum omnium fuisse ignarum? Qui illa Virgilii verba *Quid non mortalia pectora cogis, auri sacra fames* (quae quidem verba nunquam alicui vel mediocriter quidem docto dubia fuere), quid sentirent apertissime ignoravit.

[82] Cf. M. Vattasso, *Del Petrarca e di alcuni suoi amici* (Rome, 1904), 100–2. On the whole question see my *Una festa dantesca*, to be published in the near future.

[83] Cf. E. Cecchini, 'Per una nuova edizione critica dell'epistola a Cangrande', in *Miscellanea Augusto Campana*, i (Padua, 1981), 213–29.

[84] L. Bruni, 'Dialogi ad Petrum Histrum', in *Prosatori latini del Quattrocento*, ed. E. Garin (Milan-Naples, 1952), 68–70.

To begin with Dante . . . is it not clear to us that he is so often in error that he seems a complete ignoramus? He clearly did not even know the meaning—of which no one of any pretence to learning is ignorant—of Virgil's '*Quid non mortalia pectora cogis, auri sacra fames*'.

Thus, for his lack of Latin, Dante was to be removed from the ranks of the *litterati* and placed among the doublet-makers and bakers.

Mussato's impulse was not only philological and antiquarian, or rather philological and antiquarian motives were not ends in themselves. Now we must turn back to examine the great importance he attaches to the 'frame' of his vision. Mussato wanted to ensure that the poet's laurel remained an outstanding award in the cultural context of the Guelphs, a recognition of the struggle for municipal and democratic freedom against the Emperor's might. He was making this point adroitly as he sought military aid from Florence against the tyrant della Scala. He knew that he could find allies against this 'new pretender' because of Florentine spite. The Florentine Guelphs would have rejoiced in his short poem, the work of an intellectual of great prestige who, while denigrating their hated fellow-citizen, praised their noble city. Dante having become a little too famous, Mussato could provide a justification for their enmity. And Antonio degli Orsi also would have rejoiced, as he was a political enemy of Dante and more than likely a personal enemy as well. It is enough to recall that when Dante wrote his Sixth Epistle to the treacherous Florentines, accusing them of transgressing human and divine law by rebelling against the Emperor and predicting their rightful punishment—no less than in the Seventh Epistle exhorting the Emperor to tear out the roots of this 'hydra pestifera'—Antonio had vacated the bishopric of Fiesole to become Bishop of Florence. On 19 September 1312, exactly seven years before the appearance of the *Somnium*, the Emperor had effectively taken arms against the city with the aim of taming the hydra. Dante was unwilling to be among those besieging Florence, but his words—far mightier than arms—were there in spirit. On the other side, the 'right' side and the side of history, as Giovanni Villani informs us, Antonio organized the resistance (*Cronica* 9. 47): 'e 'l vescovo di Firenze co' cavalli de' chierici s'armò e trasse alla difensione della porta di santo Ambrogio e de' fossi, e tutto il popolo a piede con lui . . .' Dante 'enemy of the fatherland', stands against Antonio, 'hero of liberty'. To depict a political adversary as devoid of grammar is not without its point.

Were we to deny that the *Somnium* is anti-Dantesque, we should have to dismiss the whole fabric of allusions, borrowings, evidence of the chronology, parallels, antitheses, and contrapositions as mere coincidences. As ever in Mussato, literature has a political significance.[85] It is hardly necessary to say that Dante was not a loner, aloof, surly, and uncommitted because this was to his liking and by intellectual choice. He was, rather, vanquished and condemned to solitude by his victorious adversaries.

It has to be admitted that this lengthy consideration of the *Somnium* has involved a discussion of mediocre literature. This admission will perhaps give the excuse for a final, marginal observation on the problem raised at the beginning of this paper. It is not enough to take part in progressive movements or to have good and progressive ideas to produce good literature. Good ideas can lead to Mussato's Hell. But I should not wish to end so severely. Mussato was truly a great man. He has played, I suspect, simply a treacherously ambiguous game. It is true that he several times proclaims the truth of his vision. In the end, however, turning to the Holy Spirit, he begs for pardon, 'si quid adest aliter', 'if anything is other than true' (276). This recalls the statement of Macrobius that dreams, when the body is full of wine and food, are more like *insomnia* and lead to false revelations (*Commentarii in Somnium Scipionis* 1. 3. 1–6). Perhaps, then, all Mussato's *vera mathesis* was deceptive illusion. It should be remembered furthermore that, like Aeneas, Mussato's dove left Hades by the Gates of Ivory, those from which the *manes* send false dreams to mankind.

[85] The political use of the transcendental vision is by no means an invention by Mussato, but belongs to a pre-existing early medieval tradition: on this see W. Levison, 'Die Politik in den Jenseitsvisionen des frühen Mittelalters', in *Festgabe Friedrich von Bezold* (Bonn, 1921), 81–100 (later in his *Aus rheinischer und fränkischer Frühzeit* (Düsseldorf, 1948), 229–46); Ciccarese (n. 1 above), 390–401.

8 Literary Classicism and Latin Erotic Poetry of the Twelfth Century and the Renaissance

PETER GODMAN

> A style, when it is no longer the natural mode of expression, gains a new life—a shadowy life-in-death—as a prolongation of the past. We imagine ourselves able to revive the past through its art, to perpetuate it by continuing to work within its conventions. For this illusion of reliving history, the style must be prevented from becoming truly alive once again. The conventions must remain conventional, the forms lose their original significance in order to take on their new responsibility of evoking the past. This process of ossification is a guarantee of respectability.
>
> (Charles Rosen, *The Classical Style*)[1]

The phenomenon described by Charles Rosen in his study of Mozart, Haydn, and Beethoven demands attention in any account of the classics and post-classical Latin literature. For this historian and critic of music has evoked, with rare penetration, some of the central features of literary classicism. 'The classical style', continues Rosen, 'could originally bring no such assurance: *Don Giovanni* and the *Eroica* were scandalous, the *London* symphonies sublimely impertinent.' Rosen refers to the 'incidental . . . evocation of the past' in Mendelssohn and Schumann; to 'the sense of the irrecoverable past' in the music of Brahms; and to the irony of Mahler, concluding that 'the true inheritors of the classical style were not those who maintained its traditions, but those . . . who preserved its freedom as they gradually altered and finally destroyed the musical language which had made the creation of the style possible'.

[1] *The Classical Style: Haydn, Mozart, Beethoven* (London, 1972), 460.

I have cited at length from Rosen's conclusion not only because of the brilliance in which it is couched but also because it seems to me to reveal an insight which few literary scholars have expressed. Latin, during most of the period with which we are dealing, was not 'the natural mode of expression', and some of the most distinguished explorers of this vast territory have therefore been predisposed to ignore or to decry what Rosen has perceived and illuminated. I think particularly, in this connection, of the work of Erich Auerbach—motivated by a background in Romance philology, by the linguistic idealism of Karl Vossler, and by a Crocean reading of Vico.[2] In his interpretations of literary language, and in his analyses of post-classical literary realism, Auerbach identifies the Hegelian illusions of the *Zeitgeist* and the *Volksgeist* with the emergent voice of the people. Latin literature earns his praise in so far as it seems to prefigure the vernacular; when it evokes the past, especially the classical past, its 'ossification' is a mark of its failure. But these are not the only reasons for Auerbach's hostility to Latin literary classicism. On this subject Auerbach, like all of us, was decisively influenced by his reaction to, and against, the *magnum nomen* of Ernst Robert Curtius.[3]

No broader study of what Rosen calls 'the prolongation of the past' during our period has been written than *European Literature and the Latin Middle Ages*. The manifold conventions, the protean *topoi* of a huge literary heritage are examined by Curtius with incomparable range and intensity; and it is a tribute to the challenge of his teaching that it has stimulated a corrective

[2] Auerbach's own account of his intellectual position is given in the preface to his *Literary Language and its Public in Late Latin Antiquity and in the Middle Ages* (tr. R. Mannheim, Princeton, 1965), 3–24, and in his 'Epilegomena zu *Mimesis*', *Romanische Forschungen*, 65 (1954), 1–18 (reprinted in id., *Gesammelte Aufsätze zur romanischen Philologie*, ed. F. Schalk, Berne/Munich, 1967).

Among the many discussions of his work the fullest are those of K. Gronau, *Literarische Form und gesellschaftliche Entwicklung: Erich Auerbachs Beitrag zur Theorie und Methodologie der Literaturgeschichte* (Königstein, 1979), and G. Green, *Literary Criticism and the Structures of History: Erich Auerbach and Leo Spitzer* (Lincoln, Neb./London, 1982); and the most succinct those of Schalk in his preface to the *Gesammelte Aufsätze*, 7–18, of W. G. Fleischmann, 'Erich Auerbach's Critical Theory and Practice: An Assessment', *Modern Language Notes*, 81 (1966), 535–41, and of A. J. Evans, 'Erich Auerbach as European Critic', *Romance Philology*, 25 (1971–2), 193 ff. Valuable on Auerbach's American work and influence are H. Levin, 'Two *Romanisten* in America: Spitzer and Auerbach', *Perspectives in American History*, 2 (1968), 467–84, and P. Valegio in his foreword to the reprint of Auerbach's *Scenes from the Drama of European Literature* (Manchester, 1984). Further bibliography and analysis will be presented in my 'Auerbach and the Language of Demons' (forthcoming).

[3] Cf. P. Godman, 'The Ideas of Ernst Robert Curtius and the Genesis of *ELLMA*', Epilogue to the reprint of *European Literature and the Latin Middle Ages* (Princeton, 1990).

emphasis upon individuality.[4] Yet in the debate between Curtius and his critics both sides have lost sight of one of the fundamental paradoxes of literary classicism which Rosen has distinguished so clearly. I refer to the paradox of creative subversiveness through fidelity to the classical tradition's original attribute of freedom.

That is my theme in the present paper. My focus is on the erotic poetry written by two of the boldest interpreters of the classics in the twelfth and sixteenth centuries, Peter of Blois and Johannes Secundus. Both these poets, I shall suggest, wrote in the learned language not merely as conservative men of letters who wished to display their virtuosity to a cultivated élite but as *avant-garde* artists to whom Latin offered a uniquely sophisticated medium that they refashioned in order to formulate new cultural styles. Both chose the erotic mode not simply for private purposes of self-expression but because it presented them with specific opportunities to revive the freedom, the scandalousness, and the impertinence of the classical traditions of which they were the beneficiaries. And if neo-romantic criteria are inadequate to characterize their subversive art, no safer guide is provided by the normative prescriptions of rhetoric. For Peter, like Johannes, worked less in conformity with the school traditions of his times than in open antagonism to them, asserting, with justified confidence, the primacy of the poet as an exegete of classical poetry. Comparable in these several and significant respects, Peter and Johannes differ in another. The one wrote without drawing on the model of Catullus, probably without ever being able to read his works. The other is perhaps the most gifted Catullan imitator in neo-Latin literature. What difference, in this sphere, did the reception of Catullus make? What was possible in the erotic poetry of the sixteenth century that was not possible in the twelfth?

In the twelfth century, as in later periods, change, by a familiar irony, is often most loudly proclaimed by conservatives. There are few places in the rhetoric of *renovatio* where their voices are raised more stridently, or less convincingly, than in the *Ars versificatoria* of Matthew of Vendôme:[5]

[4] See P. Dronke, *Poetic Individuality in the Middle Ages: New Departures in Poetry, 1100–1150*² (London, 1986), with discussion and bibliography of recent works on *Toposforschung*, pp. xii ff. New perspectives on that subject are opened by L. Bornscheuer, 'Neue Dimensionen und Desiderata der Topik-Forschung', *Mittellateinisches Jahrbuch*, 22 (1987), 2 ff., and above all by Peter von Moos, *Geschichte als Topik: Das rhetorische Exemplum von der Antike zur Neuzeit und die 'historiae' im 'Policraticus' Johannes von Salisbury* (Hildesheim/Zürich/New York, 1988).

[5] *Ars versificatoria* 4. 5. The older edition by E. Faral, *Les Arts poétiques du XII*ᵉ *et du XIII*ᵉ

Antiquis siquidem incumbebat materiam protelare quibusdam diversiculis et collateralibus sententiis, ut materiae penuria poetico figmento plenius exuberans in artificiosum luxuriaret incrementum. Hoc autem modernis non licet. Vetera enim cessavere novis supervenientibus.

For ancient authors it was essential to expand their subject-matter with minor digressions and epigrams tangentially related to their topic, in order that the poverty of their material might be padded out with poetic fantasies and swell into an artificial luxuriance. For modern writers, however, this is prohibited. Overtaken by modernity, old practices have fallen by the wayside.

With the progress of research into medieval rhetoric,[6] and with the completion of Professor Munari's excellent edition,[7] it has become possible to gain a juster appreciation of Matthew and his context, and I have often asked myself, while reading stirring sentences like these, what light is cast upon the practice of an imaginative writer such as Peter by the prescriptions of a rhetorician such as Matthew.

Obvious though this question is, it is not easy to answer.[8] In many cases, the canon of important poets, such as Peter of Blois, is still provisional and debatable. And beside problems of authorship and attribution, several issues of interpretation have not yet been clarified. These are conveniently introduced by Matthew's manifesto, even though the innovations which he so peremptorily proclaims are most strikingly evinced in the Latin

siècle: *Recherches et documents sur la technique littéraire du Moyen Age* (Paris, repr. 1971), 181, is now superseded by that of Franco Munari, *Mathei Vindocinensis Opera* (Storia e Letteratura: Raccolta di Studi e Testi, 171; Rome, 1988), 195 (cf. below, n. 7). On this passage see E. R. Curtius, *European Literature and the Latin Middle Ages* (tr. W. Trask, repr. London, 1990), 490; J. Martin, 'Classicism and Style in Latin Literature', in *Renaissance and Renewal in the Twelfth Century*, ed. R. Benson and G. Constable with C. D. Lanham (Oxford, 1982), 565–6, and E. Gössmann, *Antiqui und Moderni im Mittelalter*: *Eine geschichtliche Standortbestimmung* (Veröffentlichungen des Grabmann-Instituts zur Erforschung der mittelalterlichen Theologie und Philosophie NS 23, Munich, 1974), 96–7. All translations, unless otherwise specified, are my own.

[6] For a recent survey see B. Vickers, *In Defence of Rhetoric* (Oxford, 1988), 214–53. The best analysis and critique of the use (and misuse) of rhetorical models of literary explanation are by Peter von Moos, *Geschichte als Topik* (n. 4 above) and 'Das argumentative exemplum und die 'Wächserne Nase' der Autorität im Mittelalter', in *Exemplum et Similitudo*: *Alexander the Great and Other Heroes as Points of Reference in Medieval Literature*, ed. W. J. Aerts and M. Gosman (Groningen, 1988), 55–83, with rich bibliography.

[7] Volumes i and ii of his *Mathei Vindocinensis Opera* have appeared as nos. 144 and 152 of the series Storia e Letteratura: Raccolta di Studi e Testi (Rome, 1977 and 1982). For vol. iii see n. 5 above.

[8] For a preliminary attempt see M. C. Uhl, *The Learned Lyrics of Hugh Primas*: *Interpretations of Hugh's Secular Poems with Special Reference to the 'Ars Versificatoria' of Matthew of Vendôme* (Cornell University, Ph.D. dissertation, 1977).

erotic poetry of the twelfth century by authors who eschewed his recommended medium. Less in quantitative than in rhythmical verse a fresh striving for economy, point, and elegance accompanies the reworking of inherited themes. Nowhere is this spirit of experimentation more interestingly attested than in the *œuvre* of Peter of Blois, which has been newly surveyed by Rolf Lenzen and Peter Dronke.[9] Sensitive to the example of the past and attentive to the writing of his contemporaries, Peter of Blois produced a corpus of erotic poetry that amply justifies Matthew's claims but deliberately rejects his methods. Peter's classicism is displayed in its most novel and subversive forms by two major but neglected poems preserved both in the *Codex Buranus* and in the Arundel lyric collection: *Grates ago Veneri* and *Sevit aure spiritus*.

1. Sevit aure spiritus,[10] the idealized nude and the myth of the lover

1 Sevit aure spiritus
et arborum
come fluunt penitus
vi frigorum;
silent cantus nemorum.
nunc torpescit vere solo
fervens amor pecorum—
semper amans sequi nolo
novas vices temporum
bestiali more.
Refr. Quam dulcia
stipendia
et gaudia

Wild whistles the wind
and from the trees
the leaves cascade down
in the intense chill;
silence falls on the song-filled glades.
Now the beasts grow sluggish—
their lust is afire only in springtime—
but I, the eternal lover, refuse
to follow the ceaseless changes
of the seasons like a beast.
Refr. What
sweet rewards
and happy joys

[9] See R. W. Lenzen, *Überlieferungsgeschichtliche- und Verfasseruntersuchungen zur lateinischen Liebesdichtung Frankreichs im Hochmittelalter* (diss. phil. Bonn, 1973) and P. Dronke, 'Peter of Blois and Poetry at the Court of Henry II', in *Mediaeval Studies*, 28 (1976), 185–235) (= id. *The Medieval Poet and His World* (Storia e Letteratura: Raccolta di Studi e Testi, 164; Rome, 1984), 281–339); on that of Hugh Primas see the edition by C. J. McDonough, *The Oxford Poems of Hugh Primas and the Arundel Lyrics* (Toronto Medieval Latin Texts; Toronto, 1984), with the bibliography on pp. 19–22.

[10] Texts: *Carmina Burana*, i. 2^2 ed. A. Hilka and O. Schumann (Heidelberg, 1971), 66–7; ed. B. K. Vollmann (Bibliothek des Mittelalters 13, Frankfurt, 1987), 288–291; McDonough, *Arundel Lyrics* (n. 9 above), 86–7; W. Meyer, *Die Arundel Sammlung mittellateinischer Lieder* (*Abhandlungen der königlichen Gesellschaft der Wissenschaften Göttingen*, phil.-hist. Klass. NF ii. 2; Berlin, 1908; repr. Darmstadt, 1970). Attribution: Dronke, *Medieval Poet* (n. 9 above), 331, no. 43.

Analysis of Peter of Blois' lyric forms is provided by D. Schaller, *Mittellateinisches Jahrbuch*, 14 (1979), 9–21, especially 19–20. See too McDonough, 14, and Dronke, *Medieval Poet*, no. 43. An English version of this work appears in Fleur Adcock's *the Virgin and the Nightingale: Medieval Latin Poems* (Newcastle upon Tyne, 1983), 60–5.

felicia
sunt hec hore
nostre Flore!

are these hours
spent with
my Flora!

2 Nec de longo conqueror
obsequio:
nobili remuneror
stipendio;
leto letor premio.
dum salutat me loquaci
flora supercilio,
mente satis non capaci
gaudio concipio;
glorior labore.
Refr. Quam dulcia . . .

About the long time I've spent
in Love's service I do not complain:
I am rewarded
with a noble prize;
I rejoice in a bounty of joy.
When Flora raises her eyebrows
in eloquent greeting
I cannot contain
the happiness I imagine;
I revel in my hard-pressed courtship.
Refr. What sweet rewards . . .

3 Michi sor obsequitur
non aspera.
dum secreta luditur
in camera,
favet Venus prospera.
nudam fovet Floram lectus;
caro candet tenera;
virginale lucet pectus;
parum surgunt ubera
modico tumore.
Refr. Quam dulcia . . .

I am the master of my fate,
no harsh one.
While we sport
in a secluded chamber,
Venus smiles benignly upon us.
A nude, framed by the protective bed,
Flora lies there, her skin soft and white,
her breasts suffused with the glow
of innocence, swelling just a little,
beginning to bud.
Refr. What sweet rewards . . .

4 Hominem transgredior
et superum
sublimari glorior
ad numerum
sinum tractans tenerum;
cursu vago dum beata
manus it et uberum
regionem pervagata
descendit ad uterum
tactu leviore.
Refr. Quam dulcia . . .

I become superhuman,
and delight
in being elevated
to the ranks of the gods,
as I caress her tender breasts;
divinely inspired, my hand
follows its circuitous course
and wanders past her bosom,
reaching down to her belly
with the lightest of touches.
Refr. What sweet rewards . . .

5 A tenello tenera
pectusculo
distenduntur latera
pro modulo;
caro carens scrupulo
levem tactum non offendit.
gracilis sub cingulo
umbilicum preextendit

From her soft little bosom
her slim sides taper down
in delicate proportions;
their flawless flesh
yielding
to the gentle caress.
Slender beneath the waistline
her navel stretches out

	paululum ventriculo	on a tiny belly swelling forth
	tumescentiore.	to just the slightest degree.
	Refr. Quam dulcia . . .	*Refr. What sweet rewards . . .*
6	Vota blando stimulat	My desire is aroused
	lenimine	by the alluring softness
	pubes, que vix pullulat	of the fine downy hair
	in virgine	that grows sparsely
	tenui lanugine.	on her innocent body.
	crus vestitum moderata	Her soft legs are covered
	tenerum pinguedine	with just the right amount of flesh,
	levigatur occultata	smoothly supported by a hidden
	nervorum compagine,	network of muscles,
	radians candore.	radiantly white.
	Refr. Quam dulcia . . .	*Refr. What sweet rewards . . .*
7	O si forte Iupiter	Oh, if Jupiter should chance
	hanc videat,	to see her,
	timeo, ne pariter	I fear that he too
	incaleat	would be inflamed
	et ad fraudes redeat:	and revert to his old tricks:
	si vel Danes pluens aurum	raining down to seduce her,
	imbre dulci mulceat,	like Danae, in a sweet shower of gold,
	vel Europes intret taurum,	or assuming the shape of Europa's bull
	vel Ledeo candeat	or once again becoming
	rursus in olore.	the brilliant white swan of Leda.
	Refr. Quam dulcia . . .	*Refr. What sweet rewards . . .*

The onomatopoeic opening introduces a stanza of self-conscious poise and subtlety. Although the lover in winter is a familiar figure in the verse of his contemporaries, Peter does not employ it simply in order to reject the harmony of man and nature conventionally celebrated in many twelfth-century lyrics,[11] nor is he content to draw the customary contrast between the ardour and constancy of his affections and the bleak changes wrought by hivernal chill:[12] instead, he presents subordination to the seasons as an inferior attribute of brute beasts, exalting his own sentiments to a higher and more durable plane than the transitory attachments inspired by Spring. Sempiternally passionate, the lover of stanza I expresses an exuberant confidence that is lightly developed in the refrain.

[11] Cf. W. Offermanns, *Die Wirkung Ovids auf die literarische Sprache der lateinischen Liebesdichtung des 11. und 12. Jahrhunderts* (Wuppertal/Ratingen/Düsseldorf, 1970), 115 ff.

[12] The parallels (Arundel 6. 2 f., Walter of Châtillon, Omer 18 refr. 21. 2; and *Carmen Buranum* 69. 2. 1) cited by Lenzen (n. 9 above), 78, while valid at a stylistic level, take no account of the differences of purpose and function between these two lyrics.

Hence the refusal, in stanza 2, to play the role of the plaintive suitor. The originally Ovidian vocabulary of amatory service—*obsequium* (l. 2), *stipendium* (l. 4), *praemium* (l. 5)—is employed to affirm, in the emphatic paronomasia of l. 5, joyous success.[13] Yet there is ambiguity amid the elation. The *gaudia* which Flora inspires are ineffable, passing the imagination of the awe-struck poet (2. 6–10). If the stanzas that follow (3–6) describe her body in erotic detail, erotic fulfilment is only hinted at (3. 3; 4. 5–10; 5. 6). Emphasis is initially placed upon the contemplation of idealized beauty rather than physical consummation, upon sensual reverie rather than sexual action. Flora, the perfect but passive nude, is viewed exclusively through the imagination of her exultant lover.

It is an imagination mediated by a new kind of myth. Upon the intimacies of the *chambre séparée* Venus smiles benevolently (3. 3–5). Unlike the troublesome Cupid, who reduces gods and mortals to a state of amorous agitation, Venus, in the erotic mythology of Peter's lyrics, is above all the tutelary deity of sensual delight.[14] Yet her role in *Sevit aure spiritus is* carefully limited. Neither by the well-worn device of Venus' inspiration nor through an equally conventional sense of heavenly power created by the lady's reciprocated feeling[15] does the lover attain to a state of divine exaltation (4. 1–4). His apotheosis is achieved in a tender caress (4. 5 ff.). It is the tangible nakedness of Flora that incites Peter's mythic aspirations, linking, in a manner unparalleled in the twelfth-century Latin or vernacular lyric, glorification of the lover with idealization of the nude.

That ideal is delineated on a small scale, and with a distinctive selectivity. *Cap-à-pied* description is eschewed: only the barest hint is given, at 2. 5–6, of Flora's facial features. Instead attention is focused upon her erogenous zones. The initial, discreet depiction of the maiden's budding breasts (3. 8–10) is developed, in the fifth stanza, by a number of elegant symmetries. The paronomasia of *tenello tenera* (l. 1) and of *caro carens* (l. 5) frames the matching diminutives *pectusculo* (l. 2) and *modulo* (l. 4), contributing to the sense of harmonious balance that is one of Flora's chief physical

[13] On the Ovidian *militia amoris* in medieval Latin poetry, see F. Pejenaute, *Helmantica*, 29 (1978), 195–203.

[14] See Arundel 1, especially stanzas 5 and 6 (McDonough, 74–6); Arundel 3, 1–7 (ibid. 79); Arundel 7. 2. 5–8 (ibid. 84); Arundel 10. 16 (ibid. 89) and *Veneris prosperis*, ed. P. Dronke, *Medieval Latin and the Rise of the European Love-Lyric*, ii^2 (Oxford, 1968), 393–4.

[15] On the convention see Dronke, ibid. i^2, especially 180, 279 ff.

charms. The movement of the caressing hand at 5. 6 ff. (again anticipated by 4. 6 ff.) culminates in an oxymoron of delicate virtuosity: the diminutive *ventriculo* (reinforced by the adverb *paululum*) counterpointed by the comparative *tumescentiore*, used in what Matthew of Vendôme calls its absolute sense.[16] In its intense, even mannered, refinement, Peter's portrayal of the nakedness of Flora has all the elements of a medieval 'Alexandrianism'.

It is an 'Alexandrianism' tempered, however, by wit. Sensual but innocent (3. 8; 5. 1; 6. 4), delicate but strong, Flora is described in a richly analogical style that never lapses into preciosity. Special attention is paid, for example, to her navel (5. 8 ff.), the *umbilicus* in which Isidore of Seville and the medical tradition of the Middle Ages that drew its inspiration from him located the feminine *libido*.[17] This hint of his otherwise passive mistress's sexuality is matched by her lover's arousal (6. 1 ff.) at her *pubes*, with its etymological associations of puberty and procreation.[18] And her *crus* . . . *tenerum*, covered by just the right amount of flesh, is supported by a network of hidden muscles—a physiognomical detail that lends the semi-divine perfection described by Peter in visual and tactile terms an aura of human robustness.

The humour of the final stanza emerges clearly in this context. To the lover, elevated by the sight and touch of his lady to the ranks of the gods, only Jupiter seems a fitting rival. But Jupiter, in the ardour of his competing passion, must resort to the deceptions of his fabled metamorphoses, two of which (7. 8–10) involve his assuming the shape of a bird or a beast—*bestiali more* in the disdainful expression with which the first stanza ends. Jupiter may become as white as Leda's swan, but the verbal reminiscence ('candeat rursus in olore', 7. 9–10) of 6. 10 ('radians candore') only emphasizes the superior radiance of his would-be mistress. Before the reality of her physical loveliness the topos of Jupiter's imagined mutability is reduced to a comic cliché.

Prescriptive rhetoric explains little about this witty devaluation of Jupiter. The metamorphoses and rivalry of the king of the gods had long been conventional sources of amatory comparison, codified by rhetoricians such as Matthew of Vendôme,[19] but in

[16] *Ars versificatoria* 2. 32 and cf. 27–31 (Munari, 152).

[17] *Etymologiae* 11. 1. 98, with Danielle Jacquart and Claude Thomasset, *Sexuality and Medicine in the Middle Ages* (tr. M. Adamson, Oxford, 1988), 13 and 138.

[18] Jacquart and Thomasset, ibid. 13.

[19] *Ars versificatoria* 1. 40 (Munari, 60).

the work of his more gifted contemporaries they were already becoming an object of mirth and a subject of light-hearted parody.[20] To the outmoded legend of the supreme seducer, Peter reacted, both in his letters[21] and in *Sevit aure spiritus*, both by exposing the comic potential of what was, by the later twelfth century, a literary commonplace, and by erecting a new myth in its place: the myth of the eternal lover, rapt in contemplation of the idealized nude.

'The nude is not the subject of art, but a form of art.'[22] That form is revived in *Sevit aure spiritus* in a manner that departs radically from earlier medieval traditions which associated nudity with moral turpitude or employed it to depict allegorical personifications.[23] Yet Peter's portrayal of the naked body of Flora is more than an exaltation of physical desire: it reflects an awakening of interest in the erotic representation of the idealized nude and in its aesthetic implications. Where else is that interest attested?

The fine art produced by Peter's contemporaries offers only fleeting glimpses of the eroticism fully displayed in *Sevit aure spiritus*. In the Eve sculpted by Gislebertus on the lintel fragment from Saint-Lazare at Autun, for example, a suggestion of sexuality in the prominently displayed breasts is qualified by concealment of the very erogenous zones which Peter describes so elegantly.[24] Recumbent and relaxed, his Flora assumes a position of sensual ease that is remote from the unnatural contortions of Eve's ornamental horizontality.[25] Sicily, where Peter served between 1167 and 1168 as the *praeceptor* of William II of Palermo, offers closer parallels in the celebrated nudes among the sculptural

[20] cf. the anonymous *Jupiter and Danae*, ed. W. Wattenbach, *Zeitschrift für deutsches Altertum*, 18 (1875), 457–60, and Vitalis of Blois, *Geta* 23 ff. (ed. A. Paeske, *Der 'Geta' des Vitalis von Blois* (diss. phil. Cologne, 1976), 134.

[21] For the devaluation of Jupiter cf. Peter's *Ep.* 77 (PL 207. 237–45), with comments by Dronke, *Medieval Poet* (n. 9 above), 295. Note too the contrast with *Veneris prosperis* (n. 14 above), where Jupiter's authority is invoked light-heartedly for a life of licence. On this conventional hypothetical comparison with Jupiter's amours see Lenzen (n. 9 above), 20 ff.

[22] Kenneth Clark, *The Nude: A Study of Ideal Art* (London, repr. 1973), 3.

[23] See, recently, N. Himmelmann, *Ideale Nacktheit* (Abhandlungen der Rheinisch-Westfälischen Akademie der Wissenschaften, 73; Opladen, 1985), especially 42 ff. (with full bibliography).

[24] See O. K. Werckmeister, 'The Lintel Fragment representing Eve from Saint-Lazare, Autun', *Journal of the Warburg and Courtauld Institutes*, 35 (1972), 1–20, especially 28–9, and D. Jalabert, 'L'Ève de la cathédrale d'Autun: Sa place dans l'histoire de la sculpture romaine', *Gazette des Beaux-Arts*, 6me pér., 25 (1949), 247 ff.

[25] W. Sauerländer in *Studien zur Geschichte der europaïschen Plastik: Festschrift für T. Müller* (Munich, 1965), 24–5.

decoration of the cloister of Monreale,[26] but even there the resemblances are not close. And if, in the absence of secular analogies, sacred iconography provides little with which the nude of *Sevit aure spiritus* can be compared, should we look instead to what has been called the rhetorical 'tradition' of ideals of feminine beauty in the Middle Ages[27] which allegedly enumerated the prescribed features of a perfect physiognomy and supposedly guaranteed authors' 'conformity' to 'uniform non-realistic' standards?[28]

That conventions of *descriptio pulchritudinis* were established in medieval Latin poetry of love, praise, and friendship from an early date[29] and are reflected in a range of post-Carolingian writers from Marbod to the authors of the *Carmina Cantabrigiensia* and the *Carmina Burana* is well known. But that does not demonstrate the existence of a rhetorical *tradition* of prescriptive writing on this subject before Peter's lifetime. Nor can tracts in which *descriptiones pulchritudinis* appear, such as the *Ars versificatoria* by Matthew of Vendôme, be regarded as source-books for rhythmical verse. At most, they reveal an interest in themes, forms, and styles of description which engaged Peter and other Latin poets of his age; an interest that provides not an explanation but a context for writing that transcends its prescriptions.

Both Peter of Blois and Matthew of Vendôme were pupils of Bernardus Silvestris.[30] Products of an 'intellectual milieu which valued verbal virtuosity and did not shy away from risqué themes',[31] Peter and Matthew went on to pursue contrary artistic courses. Despite his virtuous protestations that he had abandoned the (?amatory) *versus et ludicra* of his youth,[32] Peter of Blois continued, late into middle age, to compose erotic poetry. Lyrics such as the *Vacillantis trutine* (*Carmen Buranum* 108), which have been attributed

[26] See R. Salvini, *Il Chiostro di Monreale e la scultura romanica in Sicilia* (Palermo, 1962) and C. D. Sheppard Jr., *Art Bulletin*, 31 (1949), 159–69 and 34 (1952), 35–41. The wider artistic context is analysed by E. Kitzinger, 'The Arts as Aspects of a Renaissance: Rome and Italy', in *Renaissance and Renewal* (n. 5 above), 637–70, especially 666 ff.

[27] D. S. Brewer, 'Ideas of Feminine Beauty in Medieval Literature', *Modern Language Review*, 50 (1955), 157–69.

[28] Ibid. 158.

[29] See the brief discussions by Dronke, *Medieval Latin* (n. 14 above), i. 192 ff., and H. Brinkmann, *Geschichte der lateinischen Liebesdichtung im Mittelalter* (Halle, 1925), 88 ff.

[30] On Bernard and Matthew see A. Vernet, *Positions des thèses de l'Ecole Nationale des Chartes* (1937), 168 (summarized by P. Dronke, *Bernardus Silvestris: Cosmographia* (Leiden, 1978), 2). On Peter and Bernard see Dronke, ibid. 9.

[31] Ibid. 12.

[32] Ibid. 12. Peter's birth-date is unknown. The traditional speculation is *c.*1135.

to him, can be dated *c.*1190—a decade later than poems of 'repentance' like *Dum iuventus floruit* (*Carmen Buranum* 30).[33] To subscribe to a crude chronology that places Peter's erotic poetry in the 1150s and his religious verse thirty years later on the grounds that, in moralistic mood, he claimed to have renounced his *folies de jeunesse* is to ignore not only the contrary evidence of his datable lyrics but also the complexity of his artistic *persona*.[34] A poem such as *Sevit aure spiritus*, which has been assigned to 1155–70,[35] may well have been written *c.*1190, if its formal and thematic sophistication and its analogies with Peter's demonstrably later lyrics are to be trusted. We must therefore allow for the possibility that, when he wrote *Sevit aure spiritus*, Peter of Blois knew the *Ars versificatoria* (traditionally if insecurely dated to *c.*1175) by Matthew of Vendôme all too well; and we must reckon with the probability that both these pupils of Bernard Silvestris had learnt from their master the ancient lesson that the *descriptio pulchritudinis* presented a challenge to poetic virtuosity that each of them went on to meet in different ways which it is now opportune to compare.

In his *descriptio Helenae* at *Ars versificatoria* 2. 56.[36] Matthew provides a synthesis of the attributes of ideal beauty. His style, ornate and exuberant, emphasizes with inadvertent irony the natural restraint of Helen's physical perfection. No artificial embellishment enhances her loveliness; no knot confines her flowing hair. Cosmetics contribute nothing to the rosy-whiteness of her complexion; her nose is neither too long nor too *retroussé*; her lips, alluring to the kiss, hold back from a pout; and the dainty curve of her breasts refrains, as if by instinctive modesty, from a vulgar swell.

There is something comic about all this. The principle underlying Matthew's decription of Helen's beauty is *ne quid nimis*; its expression, elaborately qualified and preciously parenthetical, is manneristic. Variations on Matthew's formula are found not only in Peter's *A globo veteri* (*Carmen Buranum* 67)[37] where, in keeping with a long literary tradition, the idealized beauty is presented as

[33] On the dating of both poems see W. Lipphardt, 'Einige unbekannte Weisen zu den Carmina Burana aus der zweiten Hälfte des 12. Jahrhunderts', in *Festschrift Heinrich Besseler* (Leipzig, 1961), 101–2.

[34] Dronke, *Medieval Poet* (n. 9 above), especially 290 ff.

[35] Vollmann (n. 10 above), 1048. No explanation is given for this dating. Cf. his dating of *Grates ago Veneri* (*Carmen Buranum* 72), ibid. 1030.

[36] Munari, 82–4.

[37] Hilka-Schumann (n. 10 above), 31–2; Vollmann (n. 10 above), 222; Dronke, *Medieval Poet* (n. 9 above), 320.

Nature's masterpiece,[38] but also in the third stanza of Peter's lyric *Estas in exilium* (*Carmen Buranum* 69),[39] where the recipe is condensed with skilful economy. But *Sevit aure spiritus* corresponds to virtually none of Matthew's prescriptions. The areas of Helen's physiognomy which Matthew describes—her hair and face, her eyes, nose, and lips—are those which Peter omits entirely. With the exception of Flora's breasts, upon which Peter dwells with a loving detail absent from Matthew's catalogue, the two poets have nothing in common. Capable, as he shows in *A globo veteri* and *Estas in exilium*, of surpassing Matthew in the type of *ekphrasis* recommended at *Ars versificatoria* 2. 56, Peter of Blois consciously adopted a different style in *Sevit aure spiritus*. Was this a deliberate flouting of one prescriptive model, or did it reflect a preference for another?

Matthew himself was conscious of the ornateness, and the limitations, of his description of Helen. To the *deliciosus auditor*, the erotic connoisseur, who sets little store by wordiness, Matthew proposes an alternative model.[40] Initially succinct and compressed, it passes cursorily over the facial features that occupy most of the earlier *ekphrasis*. Matthew's attention centres instead on the region from the sides to the belly—related in terms not dissimilar to those used in *Sevit aure spiritus*.[41] Rising to its risqué culmination in the *cella pudoris* of l. 5, Matthew's description then subsides beneath a concatenation of clichés. The lady, well-proportioned and shapely, offers a pattern of moral and physical perfection. Matthew himself fancies women of this kind. For such a prize Troy was destroyed, but Paris' abduction of Helen was understandable. Put Hippolytus in his place, and he would become Priapus. . . . From the hackneyed expression of personal preference, Matthew lurches on to the debased analogy of well-worn myth, formulating no satisfactory answer to the problem of adapting poetic style to erotic subject-matter which he sensed but was incapable of resolving.

A simple solution to this dilemma lay at hand in *cap-à-pied* description that avoided the coy periphrases and boring commonplaces of the *Ars versificatoria*. Peter, we have seen, practised this type of *ekphrasis* with notable success. But its comic potential was

[38] Maria Rosa Lida de Malkiel, 'La dama como obra maestro di Dios', in id., *Estudios sobre la literatura española del siglo XV* (Madrid, 1977), 179–290.

[39] See Hilka-Schumann, 35; Vollmann, 232; Dronke, *Medieval Poet*, 322, no. 9.

[40] For text see below, pp. 178f.

[41] With *Ars* 2. 57. 3–8 cf. *Sevit aure spiritus* 5.

already apparent—parodies existed in verse possibly anterior to Matthew's set-piece on Helen[42] and in that baroque masterpiece of narrative poetry, the *Architrenius* by John of Hauville (*c.*1181).[43] There the harmony of physical proportions traditionally praised in such descriptions is burlesqued with pedantic elaboration (*omnia iusto / philosophata modo* I. 464–5), and the corporeal attributes of *moderantia* are exemplified at immoderate length and in grotesque detail. Conventions attested in earlier literature and synthesized by Matthew might inspire from his contemporaries semi-philosophical set pieces such as *A globo veteri*, polished love-lyrics such as *Estas in exilium*, or extended parodies of the *descriptio pulchritudinis* such as the one with which John of Hauville concludes Book I and begins Book 2 of the *Architrenius*, but they did not provide a viable model for erotic poetry. That Peter found for himself in *Sevit aure spiritus*, by simplifying and refashioning Matthew's flawed model.

Abandoning the comprehensiveness which Matthew recommends, Peter concentrates on achieving an erotic selectivity. In the erogenous zones, to which Matthew refers with veiled euphemism, Peter locates qualities which Matthew distributes throughout the female body. And this focus upon the sensual features of the idealized nude also magnifies the role of the suitor's aesthetic perceptions, licensing his superhuman aspirations and engendering a new poetic style—keenly visual, elegantly affective and spare—which served to refine and invigorate the coarse subjectivity and the dull mythological allusions of the *Ars versificatoria*. By a pleasing paradox, it is Peter's reaction against the *Ars* that illustrates the doctrines of innovation and economy enunciated in Matthew's treatise more successfully than Matthew's own work.[44] For his *descriptiones pulchritudinis*, verbose and conventional, are effortlessly surpassed by that *avant-garde* experiment in the erotic idiom, *Sevit aure spiritus*.

2. *Militia amoris mutata*

Weaving classical myth into a legendary fabric of its own creation, the imagination that produced *Sevit aure spiritus* proved attentive to established poetic conventions but bold in refashioning them. The

[42] Dronke, *Medieval Latin* (n. 14 above), ii. 449–52 (where the dating of the text depends on the dating of the manuscript).

[43] Ed. P. G. Schmidt, *Johannes de Hauvilla*: *Architrenius* (Munich, 1974), i. 364–487; ii. 1–74. I shall discuss this work in a forthcoming article.

[44] A further example of the literary rivalries which Matthew inspired? Cf. F. Bertini, *Sandalion*, 1 (1978), 193–206 and 2 (1979), 257–75.

same qualities are evinced in Peter's other major work of erotic poetry, *Grates ago Veneri* (*Carmen Buranum* 72)[45] where the ancient (notably Ovidian) concept of the *militia amoris* is recast in a fresh and modern mould:

1a	Grates ago Veneri,	I give thanks to Venus
	que prosperi	who with the divine favour
	michi risus numine	of her smile
	de virgine	granted me the prize
	mea gratum	I had longed
	et optatum	and hoped for
	contulit tropheum.	from my maiden.
1b	Dudum militaveram,	I'd long been in love's service
	nec poteram	without drawing
	hoc frui stipendio;	this pay;
	nunc sentio	now I feel
	me beari,	myself becoming happy
	serenari	and the clouds clearing
	vultum Dioneum.	from Dione's mien.
2a	Visu, colloquio,	My maiden had allowed me
	contactu, basio	to look at her, to speak to her,
	fruit virgo dederat;	to caress her, and kiss her
	sed aberat	but lacking
	linea posterior	was the final
	et melior	and finest limit of love.
	amori.	Unless I cross it soon,
	quam nisi transiero,	the other favours
	de cetero	she grants
	sunt, que dantur alia	are merely fuel
	materia	to my
	furori.	fire
2b	Ad metam propero.	I hasten to my goal.
	sed fletu tenero	But my little maiden, in tears,
	mea me sollicitat,	reproaches me gently,
	dum dubitat	while she hesitates
	solvere virguncula	to unfasten
	repagula	modesty's bars.
	pudoris.	As she weeps
	flentis bibo lacrimas	I drink in
	dulcissimas;	her sweet tears,

[45] Eds. Hilka-Schumann (n. 10 above), 41–2; Vollmann (n. 10 above), 244–6; Arundel 10 (McDonough (n. 9 above), 89–91). Formal and stylistic description in Dronke, *Medieval Poet* (n. 9 above), 323, no. 14, and Lenzen (n. 9 above), *Überlieferungsgeschichtliche . . . Untersuchungen* 44–6; version in Adcock (n. 10 above), 46–51.

	sic me plus inebrio,	greater draughts of drunkenness,
	plus haurio	of intense
	fervoris.	passion.
$3a^{a}$	Delibuta lacrimis	Drenched in tears,
	oscula plus sapiunt,	kisses have more flavour,
	blandimentis intimis	enticing the imagination
	mentem plus alliciunt.	with allurements of closer intimacy.
	ergo magis capior,	And so I am more bewitched
	et acrior	and my ardour is rekindled
	vis flamme recalescit.	to a fiercer flame.
	sed dolor Coronidis	But Coronis in her grief
	se tumidis	withdraws, sobbing and sighing,
	exerit singultibus	and concedes nothing
	nec precibus	to my
	mitescit.	entreaties.
3^{b}	Preces addo precibus	I redouble prayer upon prayer,
	basiaque basiis;	kiss upon kiss,
	fletus illa fletibus,	she tears upon tears,
	iurgia conviciis,	and bickering with rebuke.
	meque cernit oculo	Now she looks at me
	nunc emulo,	with a glower of enmity,
	nunc quasi supplicanti;	now with an air of virtual pleading,
	nam nunc lite dimicat,	at one moment battling fiercely,
	nunc supplicat;	at another begging my favour,
	dumque prece blandior,	and the gentler she becomes at my
	fit surdior	prayer,
	precanti.	the deafer she is to what I pray.
4^{a}	Vim nimis audax infero.	With bravado I bring force to bear.
	hec ungue sevit aspero,	She fights back tooth and nail,
	comas vellit,	clawing at my hair,
	vim repellit	vigorously repelling
	strenua,	my onslaught,
	sese plicat	bending in two
	et intricat	and crossing her knees,
	genua,	so that the door of modesty
	ne ianua	is not
	pudoris resolvatur.	unbarred.
4^{b}	Sed tandem ultra milito,	Undaunted, I soldier on
	triumphum do proposito.	winning the victory I desire,
	per amplexus	I tighten my embrace
	firmo nexus,	fast upon her,
	brachia	pinning down her arms
	eius ligo,	and press kisses

	pressa figo	hard upon her;
	basia;	so the realm of Venus
	sic regia	is unbarred.
	Diones reseratur.	
5^{a}	Res utrique placuit,	It pleased us both,
	et me minus arguit	and my gentler lover
	mitior amasia,	rebuked me less,
	dans basia	giving honeyed kisses,
	mellita	
5^{b}	Et subridens tremulis	With a bashful smile
	semiclausis oculis,	in her trembling
	veluti sub anxio	half-closed eyes,
	suspirio	as if lulled to sleep
	sopita.	beneath a troubled sigh.

Again Peter begins by belying conventionality, introducing three inherited ideas that are freshly developed later in his poem: the favour of Venus (1^{a} and 1^{b}), the *militia amoris* (1^{a}. 7 and 1^{b}. 1–3), and the *quinque lineae amoris* (2^{a}).[46] Yet it is not simply a celebration of the act of intercourse or even a lyrical account of seduction that is his chief concern in *Grates ago Veneri*. These are the themes that he treats in slighter erotic poems, rejoicing in intimacy or resigning himself to a problematic preference for love over study.[47] Tempering the tone of triumph with which *Grates ago veneri* opens is an emphasis on the innocence of Coronis—described as *virgo* at 1^{a}. 4 and 2^{a}. 4, as *virguncula* at 2^{b}. 6. Her vulnerability is underscored by the metaphor of the door (2^{b}. 7–8, borrowed from Roman elegy and transposed into a religious context in Peter's later poetry[48]) in which her reluctance is conveyed (2^{b}. 7–8). Despite all the superficial ebullience with which the persona of the lover proclaims his determination and extols his conquest, *Grates ago Veneri* is more than a love-song 'in which a lover describes a seduction in exultant detail'.[49] It is an erotic poem of greater delicacy than any other that Peter was to write.

For the theme of loss of innocence is also treated by Peter in a

[46] For bibliography see McDonough (n. 9 above), note (90) ad loc., and Dronke, *Medieval Latin* (n. 14 above), ii. 448–9.

[47] e.g. *Ipsa vivere michi reddidit*, Arundel 3 (McDonough 79) and *Ver prope florigerum, flava Licori* (Dronke, *Medieval Latin*, ii. 374–6) with discussions of attribution by id., *Medieval Poet* (n. 9 above), 325 (n. 20) and 332 (no. 47).

[48] Arundel 20. 4a. 9–10 (McDonough, 104). Cf. n. 34 above. Cf. the echo of 1b. 1 in *Olim militaveram* (Dronke, *Medieval Poet*, 337–8).

[49] Dronke, ibid. 295.

brutal parody of Horatian themes.[50] 'Wit' and 'frivolity'[51] are not the most salient features of this lament of a twelfth-century Humbert Humbert over the precocious promiscuity of his Lice-Lolita. Beginning with an echo of Horace, *Odes* 3. 6. 24, *Ex ungue primo teneram* describes the self-righteous sorrow of a sensualist cheated of his prize. Was *Ex ungue primo teneram* written as an antithesis to the praise of qualified continence and restraint from the last *linea amoris* that is *Carmen Buranum* 88?[52] The hypothesis is speculative, but not so the points of comparison and contrast between *Grates ago Veneri* and *Ex ungue primo teneram.* Both poems treat the loss of virginity from a perspective of masculine single-mindedness—the one decrying, the other celebrating the changed state of a former *virguncula.* But where the failed lover of *Ex ungue primo teneram* tempers his egotistical complaint only with coarse humour, the successful suitor of *Grates ago Veneri* introduces a note of ambiguity into even his most confident eloquence, providing an inadvertent insight into the response of Coronis which we are denied in Lice's case. From this ambiguity *Grates ago Veneri* derives some of its subtlest effects.

Opening as a grateful invocation by the lover to his tutelary deity (1^{a-b}), *Grates ago Veneri* swiftly develops a series of contrasts. His dash to the metaphorical goal (2^{b}. 1) is couched in the literal language of the ancient race (*linea, meta*). Coronis' reluctance is transposed by the lover into an epigrammatic style that both displays the range of his eloquence and reveals the limitations of his understanding. Her tears—now gentle, now sweet (2^{b}. 2, 8)—are an intoxicating draught of passion (10–12), an aphrodisiac to her *oscula* (3^{a}. 1–4). Before his single-minded ardour her pain and sorrow (3^{a}. 5–12) produce contrary reactions which he resolves only verbally, in the witty paradox with which 3^{b} culminates (10–12).

The verbal and emotional struggle now (4^{b}) becomes a physical conflict, and the traditional amatory metaphor of the door introduced at 2^{b}. 6–7, is given a brilliant twist in Peter's risqué image of the *exclusus amator*, prevented from penetration in the most literal sense. And it is in this interplay between literal and metaphorical diction that *Grates ago Veneri* achieves much of its complexity, for the conventional vocabulary of *militia amoris* at 1^{a} and 1^{b} now

50 *Ex ungue primo teneram* (Dronke, *Medieval Latin*, ii. 378–80; id., *Medieval Poet*, 322 (no. 11)). Cf. Horace, *Odes* 3. 6. 21 ff. and 10 (Lyce).

51 Dronke, *Medieval Latin* ii. 380.

52 Hilka-Schumann (n. 10 above), 78; Vollmann (n. 10 above), 290–1.

acquires a sinister concrete aptness: the lover soldiers on (4^{b}. 1), winning his conquest by an act of force (4^{b}. 2–8) which the figurative language of unbarring and entry (4^{b}. 9–10) does nothing to conceal. At one level, we are invited to perceive this 'seduction' in the familiar terms of Ovidian analogy. At another, new doubt is engendered by the discrepancy between the polish with which these literary conventions are handled and the violence of the act which they describe.

Hence the ambiguous stress on reciprocity at 5^{a}. 1. The situation, like the tense, changes: Coronis is no longer the refractory *virguncula*, described in the historic present, of previous stanzas. She has become a *mitior amasia*, remembered in the past (5^{a}. 1–2), less inclined to cross her lover's will. The *me* of 5^{a}. 2, however, sounds the same note of egocentricity that is so insistent earlier in the poem;[53] Coronis' half-closed eyes tremble, her smile is tentative, her sight anxious (5^{b}). The 'exultant detail' in which the lover proclaims his success is undermined by the elusiveness of his conquest, and the final image is not one of reciprocated ardour but one of withdrawal into troubled sleep.[54] *Grates ago Veneri*, superficially so confident a paean of praise for the lover's triumph, closes by gently suggesting the limitations of his ability to enter the mind of his mistress. *Militia amoris* ends both in physical conquest and in imaginative defeat.

No less remarkable than the imaginative tact of this poem is the discrimination which it displays. Peter was not only capable of recasting established conventions; he was also able to resist the temptation of models that lay readily to hand. Of his intimate familiarity with Ovid's works, especially the *Amores*, there can be no doubt: Peter alludes to them time and again. An obvious Ovidian precedent for what has been thought to be Peter's purpose in *Grates ago Veneri* was available in *Amores* 1. 5; and other twelfth-century poets did not hesitate to draw on this well-known source.[55] But Peter, in *Grates ago Veneri*, was no more attempting to depict a scene of seduction than he was seeking, in *Sevit aure spiritus*, to compose a *descriptio pulchritudinis*. The simpler strategy of imitation was one he rejected in order to create something more

[53] 1^{a}. 5, 1^{b}. 5, 2^{b}. 3, 10, 3^{b}. 5.

[54] On the possibility of an Apuleian echo here see Dronke in *Renaissance and Renewal* (n. 5 above), 597.

[55] See T. Latzke, 'Die Carmina Erotica der Ripollsammlung', *Mittellateinisches Jahrbuch*, 10 (1974), no. 4, 175–6.

complex. With Ovidian boldness, the ancient amatory metaphors of *militia amoris* are invested with new literal meaning, and a note of sophisticated ambiguity is introduced into the treatment of a subject that Ovid had handled unequivocally. In this respect *Grates ago Veneri* demands to be distinguished from other twelfth-century lyrics which reflect Ovid's influence, invoke Venus, or ring the changes on the theme of the *quinque lineae amoris*. Drawing on the traditional language of amatory description, Peter radically but fastidiously alters it in a manner that both preserves its resemblance to its sources and emphasizes its differences from them. Rivalry with Ovid is evident here, but so too is the tribute to his contemporaneity. In Rosen's paradox, it is Peter's transformation of his model that serves to affirm its modernity.[56]

That paradoxical affirmation of Ovid's modernity thus entailed a claim to an equal status—the status of a 'classic', in twelfth-century terms, which Peter won by the techniques of a learned 'Alexandrianism'. No clearer examples of Warburg's celebrated aphorism—'Athen will eben immer wieder neu aus Alexandrien zurückerobert sein'[57]—exist in the Latin erotic poetry of the twelfth century than *Sevit aure spiritus* and *Grates ago Veneri*, works that depend upon an erudite tradition from which they simultaneously assert their liberty.

Through emulation with antiquity both Peter's poems lend substance to the lofty pretensions of the *moderni* among the rhetorician-contemporaries while rejecting their mundane methods. No formula of rhetorical composition accounts—except negatively—for *Sevit aure spiritus* and *Grates ago Veneri*. No system of the *artes* produced during the twelfth century makes a place for art of this subversive, shocking kind. That is why when we read, at the beginning of the long section of the *Codex Buranus* that contains its incomparably rich and still imperfectly understood collection of love-poetry, the first of several lyrics dramatizing the conflict between *Amor* and *Studium*,[58] we should perhaps hesitate before invoking the ubiquitous spectre of literary '*topoi*' or the familiar phantom of 'the goliardic *persona*'. These terms conjure up illusions that obscure a far more interesting reality—the reality of gifted medieval poets, such as Peter, seeking to forge a weapon incisive enough to sever

[56] Cf. n. 1 above, and Rosen and H. Zerner, *Romanticism and Realism*: *The Mythology of Nineteenth Century Art* (London/Boston, 1984), 174 (Titian and Manet).

[57] Aby Warburg, *Gesammelte Schriften*, ii (Leipzig/Berlin, 1932), 534, with E. Gombrich, *Aby Warburg*: *An Intellectual Biography* (London, 1970), 214–15.

[58] Hilka-Schumann (n. 10 above), 56; Vollmann (n. 10 above), 168.

themselves from the very conventions that had made their art possible.

In his choice of subject—outrageous to orthodox morality—in his manipulation of form, theme, and allusion—startlingly novel yet subtly refined—Peter of Blois was not only aiming to achieve a revival of erotic verse and characterizing his independent stance toward the authority of antiquity. He was also affirming his freedom from the constraints of a school tradition by which Matthew, its conservative continuator, remained imprisoned. Peter was not alone among his contemporaries in taking this stance, as I hope to show elsewhere,[59] but he was not to meet his match in the competitive sphere of Latin erotic poetry until the sixteenth century, when Johannes Secundus published his *Basia*.

To approach Johannes Secundus' poetry was, until recently, hardly less difficult than it was to approach that of Peter of Blois.[60] An adequate commentary on Johannes' works has never been provided; the only respectable edition is more than a century old and not always easy of access.[61] Nor were the details of Johannes' biography established in depth until the publication, in 1986, of Dr A. M. M. Dekkers's dissertation.[62] But even Dekkers, who has laid such a solid factual foundation for the study of Johannes' life and works, has shown us only one side of his *œuvre*. It is Johannes the occasional poet, the panegyrist of the Habsburg imperial family, the encomiast of alliances and treaties, who emerges from Dekkers's pages. This is a valuable, if austere, corrective to some of the sloppy romanticism that Johannes' writing has attracted in the English-speaking world, but it leaves a number of questions about his major productions in amatory verse unanswered. Without trespassing on the territory that Professor Ludwig has explored,[63] I should like to outline some of these questions and sketch possible, if provisional, answers to them.

What was the function of the erotic poetry that Johannes composed with such skill? Mere 'artificial allusive verse . . . as decorative,

[59] P. Godman, *The Silent Masters* (Princeton, forthcoming).

[60] This paragraph draws on my review-article 'Johannes Secundus and Renaissance Latin Poetry', *The Review of English Studies*, NS 39 (1985), 258–72.

[61] *Ioannis Nicolaii Secundi Hagani Opera Omnia, emendatius et cum notis adhuc ineditis Petri Burmanni Secundi denuo edita cura Petri Bosscha* (Leiden, 1821). There is a reprint of the *editio princeps* of 1544 (Nieuwkoop, 1969).

[62] *Janus Secundus (1511–36): De tekstoverlevering van het tijdens zijn leven gepubliceerde werke* (Nieuwkoop, 1986).

[63] Below, pp. 183–97.

harmless, and sophisticated as the elegant nonsense of the Neoplatonists'?[64] In the case of Peter of Blois we have seen that the artificiality and allusiveness of his medium were integral to its meaning, and that its sophistication was employed to formulate a radical reinterpretation of his chosen classical style. Johannes' work displays the same characteristics, although its origins and purpose are not easy to discern.

In Johannes' immediate circle there are no obvious precedents for the type of erotic verse represented by the *Basia* and some of the *Amores*.[65] The greatest scholar connected with the Evraerts family, Erasmus, wrote predominantly religious poetry.[66] The true precursors of Johannes, in this sphere, are Italian—chiefly Marullo, Sannazaro, and above all Pontano.[67] But neither the general European diffusion of their works nor the particular influence of Alciato upon Johannes during his sojourn at Bourges sufficiently explains why, in an age in which poetry was classified as a part or instrument of moral philosophy,[68] and a school of neo-Aristotelian thought argued that poetry should be an image or representation of the truth,[69] an author as undoubtedly versatile and as demonstrably opportunistic as Johannes chose to compose a complete cycle of lyrics in the apparently frivolous *basium*-genre. Simply to say that he wished to imitate Catullus is, in its early sixteenth-century setting, no answer at all.

That is why the weary shorthand of 'school exercises' is patently inapplicable to the amatory works of this neo-Latin poet. Johannes goes further than Peter in emphasizing not merely his differences from the *grammatici* but also his antagonism to them. He does so, most notably, in his much-anthologized epigram *Ad grammaticos* and in *Basium* 12.

Basium 12[70] begins on a note of feigned surprise voiced in a vocabularly rich in generic associations. The contrast which

[64] A. Grafton, *Joseph Scaliger: A Study in the History of Classical Scholarship*, i. *Textual Criticism and Exegesis* (Oxford, 1983), 41.

[65] See L. Verhoeven, *De invloed van Catullus op de Neolatijnse dichters uit de Nederlanden voor 1550* (diss. Katholieke Universiteit te Leuven, 1966).

[66] See C. Reedijk, *The Poems of Desiderius Erasmus* (Leiden, 1956).

[67] See Walther Ludwig, pp. 183–97 below.

[68] A useful conspectus and bibliography of views on this subject is provided by D. Robey, 'Humanist Views on the Study of Poetry in the Early Renaissance', *History of Education*, 13 (1984), 7–25.

[69] Fundamental remains B. Weinberg, *A History of Literary Criticism in the Italian Renaissance*, 2 vols. (Chicago, 1961), 6, 31 ff. See now A. Fisher, 'Three Meditations on the Destruction of Vergil's Statues: The Early Humanist Theory of Poetry', *Renaissance Quarterly*, 40 (1987), 607–35.

[70] For text see below, Appendix, p. 179.

Johannes draws between the *carmina mentulata* attributed to him and the *basiationes* he admits to composing adapts the language of Catullus 16, partly mediated through Martial.[71] Johannes' ironical profession of personal virtue (*Basium* 12. 8–9) relies on a Catullan defence (16. 5–6) of similar ambiguity. But self-justification swiftly turns into attack upon the *hirsutus magister* and the dubious innocence of his charges. Mocked for his prudery in *Basium* 12, the teacher is also ridiculed in Johannes' epigram on grammarians where, with a parody of Horatian *hauteur* (*Odes* 3. 1. 1), the desire to keep pedants at bay is adduced as a reason for writing dirty verse.[72] Both this epigram and *Basium* 12 combine an inversion of the high style of the first of the Roman *Odes* with humorous echoes of Catullus in order to assert the primacy of the poet over the *grammaticus* and affirm the independence of his verse from the classroom context. The encomiast of Charles V rejects the theme of *magnanimi . . . Caesaris arma* that can be reduced to a school-exercise, and opts for the unedifying genre of *basiationes*. Is this merely an elegant literary pose, or do deeper cultural motives lie behind it?

The *Basia* were not written in blithe indifference to the *grammatici*, despite all Johannes' satirical scorn. The poem which introduces this cycle draws upon a standard *exemplum* in Rudolph Agricola's version of Aphthonius, probably known to Johannes through Erasmus.[73] But the 'total routineness of imaginative writing'[74] which Aphthonius had sought to instil is resisted by Johannes in his variant on the familiar theme of Venus and the roses. Unlike Aphthonius, for Johannes in *Basium* 1 it is Venus' nostalgic kisses, inspired by the memory of Adonis, that make roses red. On to an idyllic setting adapted from the first book of the *Aeneid* (1. 680–94), Johannes superimposes the disconsolate passion of the goddess, his opening allusion to Virgil's epic preparing the way for his final appropriation of Latin as the special medium of erotic fantasy. And by recreating this myth to extol the genre which he practices in defiance of the *grammatici*, Johannes annexes their didactic province to his personal realm of love-poetry. *Basium* 1 is thus more than a decorative exercise in verse composition theme. It is an act of aggressive *aemulatio* directed not only at

[71] Appendix, p. 179. [72] Appendix, pp. 179 f.

[73] Discussed by A. Grafton and L. Jardine, *From Humanism to the Humanities: Education and the Liberal Arts in Fifteenth- and Sixteenth-Century Europe* (London, 1986), 128 ff., esp. 132–3. For background see P. Mack, 'Rudolph Agricola's Reading of Literature', *Journal of the Warburg and Courtauld Institutes*, 48 (1985), 23–41. For texts see below, Appendix, pp. 180 f.

[74] Grafton and Jardine (n. 73 above), 131.

the works of Virgil but also at the stock-in-trade of the *insulsi grammatici*.

Why did Johannes think so little of them? How did their writings influence his own response to classical poetry and to Catullus in particular? Some of the answers to these questions must remain conjectural, pending further investigation of Johannes' milieu and of the earliest Catullan scholarship, but I venture to think that it is possible, even in the present imperfect state of our knowledge, to hazard some partial solutions.

By the late fifteenth century, when manuscripts of Catullus had become less uncommon and his works were regularly accompanied by those of Tibullus and Propertius (as they are in the Venetian *editio princeps* of 1472), a philological tradition of commentary on Catullus' poetry was beginning to be established that lends substance to Johannes' aspersions.[75] The projected commentary of Politian, sadly pre-empted by that of Antonius Parthenius in 1485, might have refuted this assertion, as the Catullan interpretations embedded in the *Miscellanea* strongly suggest, but it was never published.[76] Between 1485 and the death of Johannes in 1536 the full-scale Catullan commentaries by Parthenius, Palladius Fuscus (1496), and Alessandro Guarino (1521), incorporating the lecture-notes of his father Baptista, are dominantly didactic in purpose—narrowly philological, generally elementary, often downright banal. The verbose deserts of exegesis that surround the small oases of text in their works cast up little more than grammatical observations, lists of synonyms, explanations of geographical and mythological references, and vague digressions. Let me illustrate from Alessandro Guarino's commentary on Catullus 5.[77]

[75] The best conspectus is that of J. Haig Gaisser in her forthcoming entry on Catullus in the *Catalogus Translationum et Commentariorum*. Little is added by L. A. Colliard, *La Fortuna in Francia del 'Liber Catulli Veronensis' nel XVI Secolo* (Verona, 1978).

[76] See Gaisser, 'Catullus and His First Interpreters: Antonius Parthenius and Angelo Poliziano' *Transactions of the American Philological Association*, 112 (1982) 83–106.

[77] Alessandro Guarino on Catullus 5 (selections from *Alexandri Guarini . . . in C. V. Catullum . . . per Baptistam patrem emendatum expositiones* (Venice, 1521)).

Catullus puellam Clodiam amavit. Sed pro Clodiam Lesbiam nominavit, sicut Apuleius in Apologiae oratione prima annotavit, similiter etiam quae Metella erat Perillam scripsit; sicut Propertius . . . *vivamus*: in hoc verbo voluptas quaedam inclusa est, ac si diceret genio indulgeamus, sicut apud Martialem . . .

atque amemus: amare est diligere, licet in illo maior quaedam vis insit. Et amare alias est amore frui, sicuti declarat Donatus apud Terentium in *Eunucho* . . .

da mihi basia mille: cum igitur ita sit quod mortui numquam reviviscunt, ab amica, dum vivit et potest voluptatibus frui, basia petit.

deinde centum . . . milia multa fecerimus: supradicta namque 3300 sunt . . .

Guarino begins by identifying Lesbia as Clodia, and by discussing the question of pseudonyms. He goes on to consider, with no great penetration, the flavour of the opening word *vivamus* and to analyse the various verbs in Latin for loving. Passing on through a list of synonyms, he comes to one of the finest instances of the numerical theme in Roman love-poetry: 'da mihi basia mille, deinde centum'. What does Guarino have to say about this? He speculates that Catullus, aware that he must go the way of all flesh, is seizing the opportunity of fleeting happiness; and, observes Guarino, if you add up the number of kisses, they come to 3,300!

One can acknowledge the relative infancy of the humanist commentary in this period[78] and concede that scholarship of this kind served a purpose, but there are limits to even historical sympathy. If one sets Guarino or his precursors in the commentary tradition beside their contemporaries, the neo-Latin poets whose imitations of Catullus amounted to nothing less than sustained and subtle Catullan interpretations, then one can gauge the standards of comparison implied by the phrase *insulsi grammatici*. An example is provided by Johannes' *Basium* 7, which I cite with the 1647 translation of Thomas Stanley:[79]

Centum basia centies,
Centum basia millies,
Mille basia millies,
Et tot milia millies,
Quot guttae Siculo mari,
Quot sunt sidera caelo,
Istis purpureis genis,
Istis turgidulis labris
Ocellisque loquaculis
Ferrem continuo impetu,
  O formosa Neaera!
Sed dum totus inhaereo
Conchatim roseis genis,
Conchatim rutilis labris
Ocellisque loquaculis,

Kisses a hundred, hundred fold,
A hundred by a thousand told,
Thousands by hundreds number'd o'er
As many thousand, thousand more
As are the drops the seas comprise,
As are the stars that paint the skies,
To this soft cheek, this speaking eye,
This swelling lip, will I apply.
  But whilst on these my kisses dwell
Close as the cockle clasp her shell,
This swelling lip I cannot spy,

[78] Succinctly discussed, with full bibliography, by J. H. Waszink in *Leiden University in the Seventeenth Century: An Exchange of Learning* (Leiden, 1975), 164 and nn. 19–21. See too A. Grafton, 'The Scholarship of Politian and its Context', *Journal of the Warburg and Courtauld Institutes*, 40 (1977), 152–6.

[79] *Kisses, being the Basia of Joannes Secundus rendered into English by Thomas Stanley* (repr. London, 1927), no. 7. Note that at l. 3 the copy-text for the 1541 edition (Oxford, Bodley MS Rawlinson 154) reads *centies*.

Non datur tua cernere
Labra, non roseas genas
Ocellosque loquaculos
  Molles nec mihi risus,
Qui, velut nigra discutit
Caelo nubila Cynthius
Pacatumque per aethera
Gemmatis in equis micat,
  Flavo lucidus orbe,
Sic nutu eminus aureo
Et meis lacrimas genis
Et curas animo meo
  Et suspiria pellunt.
Heu, quae sunt oculis meis
Nata proelia cum labris?
Ergo ego mihi vel Iovem
Rivalem potero pati?
Rivales oculi mei
  Non ferunt mea labra.

This softer cheek, this speaking eye:
Nor those sweet smiles, which (like the ray
Of Cynthius driving clouds away)
From my swoln eyes dispel all tears,
From my sad heart all jealous fears.
Alas! what discontents arise
Betwixt my oemulous lips and eyes!
Can I with patience brook that Jove
Should be a partner in my love,
When my strict eye the rivalship
Disdains to suffer of my lip?

Basium 7, like the Catullan poems 5 and 7 upon which it draws, is paired with another work by Johannes. In *Basium* 6, a stylish variation on the numerical theme decrying the possibility of counting kisses and elevating the mistress to semi-divine dignity, there is no hint of the *invidia*, the outside threat to the lovers, that Catullus adduces as the pretext for his exigency. But in *Basium* 7, its companion-piece, allusion becomes parody all the more telling as interpretation for its deeper debts to Catullus' style. His fifth and seventh poems are explicitly evoked at the opening of Johannes' *Basium* 7 in a vertiginous numerical hyperbole. Bedazzled by this luxuriance of neologism and diminutives in the Catullan manner,[80] the reader of *Basium* 7 takes time to realize that the entire catalogue of Catullan *topoi* in the first nine lines of this poem depends on the dubitative subjunctive *ferrem* in the tenth; that it is not until almost half-way through the text, with the *non* of l. 16, that the poem's apparently triumphant tone betrays something more equivocal; and that one must wait until the very end of *Basium* 7 to see that the rivals Johannes fears are aspects of his own physiognomy. In a conventional comparison, also émployed by Peter of Blois,[81] Johannes asks, what chance he would stand

[80] On this feature of Johannes' style see the exhaustive study of L. Vander Elst, *Het Latijnse diminutief en zijn gebruik door Janus Secundus* (*1511–1536*) (diss. Katholieke Universiteit te Leuven, 1969).

[81] See above, pp. 155–8.

against the competition of Jupiter? None at all, is his novel answer. He can't even resolve the rivalry between his eyes and his lips. The external *invidia* of Catullus 5 and 7 gives way to a comic introversion at the end of *Basium* 7; the insatiable suitor with his mistrust of the outside world in Johannes' model becomes the incapable suitor whose self-division produces ineptitude. Catullan themes are taken to their limit, and perhaps beyond, in an outrageous style that has few contemporary parallels. Other neo-Latin poets responded, often ingeniously, to the distinctive themes of Catullus 5 and 7; others handled them imaginatively; but none known to me was capable of suggesting, with Johannes' wit and intelligence, how Catullus could be funny. Compared with Guarino and his predecessors commenting on Catullus 5, this poem reflects not only a different activity but also a different quality of mind. From Johannes' urbane perspective, the *insulsi grammatici* were given their unflattering epithet because it was true.

It is 'the naughty Catullus', the 'gay dog', who appears most prominently in the work of his neo-Latin imitators, claims Mary Morrison in her still valuable study of this subject.[82] 'When they write about love these poets tend to become shallow, imitative, monotonous . . . all of them repeat *ad nauseam* the sentiments of Petrarch in the manner of Catullus or Propertius. Their love-lyrics are catalogues of beauties, gardens of sensual delights. Johannes Secundus found the recipe for this kind of writing . . .', concludes John Sparrow in a well-known essay.[83] It is difficult to reconcile these sweeping statements with what we find in the intricate detail of Johannes' text.

For the 'Alexandrianism', the pseudo-Catullan preciosity of *Basium* 7, incorporates an unmistakeable irony, directed partly at Johannes' model and partly at the poet himself. The mannerisms which he adapts from his sources disclose, by their exaggeration, its potential humour and project it back on Johannes' own style of imitation. And when he compares himself to his ancient precursor, as he does in *Basium* 16,[84] there is nothing frivolous or shallow about the resemblance he evokes. Swiftly passing from

[82] 'Catullus in the Neo-Latin Poetry of France before 1550' and 'Catullus and the Poetry of the Renaissance' in *Bibliothèque d'Humanisme et Renaissance*, 17 (1955), 365–94 and 25 (1963), 25–56.

[83] 'Latin Verse of the High Renaissance' in *Italian Renaissance Studies*, ed. E. F. Jacob (London, 1960), 381.

[84] For text see below, Appendix, pp. 181 f.

the numerical theme to the more sombre subject of the bitter-sweetness of love, Johannes draws on the dark imagery of Catullus 68. Death and sorrow are the themes he takes from his Catullan model, fusing them with an allusion to Ovid, *Amores* 3. 7. But it is impotence, not fulfilment, that is the subject of Ovid's poem, and the memory of this source, combined with that of Catullus, pervades *Basium* 16. Filled with yearning imperatives and hypothetical subjunctives, Johannes' work plunges from the optimistic assurance of Catullus 5 and 7 into a pessimistic recognition of the unreality of his passion. The mistress in his feverish embrace has the coldness of a corpse. Clinging to her in a swoon, he implores the kiss of life. Consciousness of the proximity of death, intimated by echoes of Catullus 68, blight the *aevi tempora floridi* of *Basium* 16, and the poem ends with a vision of disease and extinction. A 'gay dog' frisking in 'gardens of sensual delight?' I detect no evidence for these clichés in Johannes' exploitation of the ambiguity of his models.

When Julius Caesar Scaliger's canon of neo-Latin poetry appeared (posthumously) in 1561,[85] it made little room for lyric poetry and excluded Johannes Secundus.[86] On Scaliger's dominantly didactic criteria, such an exclusion is understandable, for Johannes deliberately set himself against the canonical claims of the 'higher' genres, against the prudish conventions of the schoolroom, against the banalities of routine exegesis. His *persona* is that of a rebel—but a rebel of a complex kind. Like Peter of Blois, Johannes was able to fuel the fires of his revolt within and without a highly developed literary tradition. It had mediated to him a *Catullus renatus*, the Catullus of his fifteenth-century neo-Latin imitators—epicurean, sensual, elegant to the point of mannerism.[87] And that manneristic element in Johannes' poetry, as in the art of his contemporaries, represented less a break from the constraints of a normative classicism than an alternative, 'Alexandrian', interpretation of his chosen Catullan styles.[88]

[85] Edited I. Reineke, *Julius Caesar Scaligers Kritik der neulateinischen Dichter* (Munich, 1988), discussed by W. Ludwig, *Litterae Neolatinae* (p. 183, first footnote below), 220 ff.

[86] Ludwig, 239–41.

[87] Ludwig, ibid.

[88] Critiques of W. Friedlaender's celebrated study 'Die Entstehung des antiklassischen Stils in der italienischen Malerei um 1520', *Repertorium für Kunstwissenschaft* 46 (1925), 49 ff. (reprinted, in English translation, in his *Mannerism and Anti-Mannerism in Italian Painting* (New York, 1957), 3 ff.) are neatly summarized in *Zauber der Medusa*, ed. W. Hofmann (Hamburg, 1988), 27 ff. Comparative discussion in J. V. Mirollo, *Mannerism and Renaissance Poetry: Concept, Mode, Inner Design* (New Haven and London, 1984), 1–71.

The model provided by Johannes Secundus exercised an enduring influence, both on learned and on vernacular literature, from the sixteenth to the eighteenth centuries, until it was dealt a mortal blow by Goethe in his important but neglected lyric *An den Geist des Johannes Secundus*.[89] Its popularity incurred the odium of modern critics such as Benedetto Croce, who branded Johannes and his imitators trivial dilettanti whose preciosity tended naturally to the excesses of *marinismo*.[90] 'La lingua, già adustata e resa scorevole a certe combinazioni d'imagine e di parole, poetava per suo conto in luogo del poeta', declared Croce in an adaptation of Goethe.[91] This Crocean flourish (ironical in one allegedly so hostile to rhetorical hyperbole) was aimed at form without content—at what Spitzer called the absence 'of new personal emotion in traditional Latin vocabulary';[92] at what a recent scholar likens to 'elegant nonsense'.[93] The only difference between Croce and his covert or unconscious sympathizers is that Croce was *openly* romantic and sufficiently candid to present an argument for a position which seems to me—for reasons considered in this paper—palpably false.

It is not simply that the qualities that Croce identified with the anti-intellectual strain of romantic poetry are evinced in the classical styles we have examined—less by the display of what Croce called 'energia fantastica' and 'forza vitale' than through their paradoxical fidelity to the original freedom of their models. Nor is it sufficient to register the strictly stylistic changes that the reception of Catullus made possible in the Latin erotic poetry of the early Renaissance. More important, perhaps, than either of these literary changes are their cultural implications; and it is here that Johannes' method of writing, or re-writing,[94] demands to be compared with that of Peter of Blois.

Both Peter and Johannes subvert and transform their models. Their techniques are eclectic, their imaginations mythopoeic. They are concerned to create new myths of the poet, in contrast to the prescriptions of the *grammatici*. Both Peter and Johannes were

[89] See P. Godman, '*Johannes Tertius*: Goethe and Renaissance Latin Poetry' (*Journal of the Worburg and Courtauld Institutes* 53 (1990), 101–16, forthcoming).

[90] 'Poesia popolare e poesia d'arte', in id. *Scritti di storia litteraria e politica*, xxvii (Bari, 1933), 463–4.

[91] Ibid. 464.

[92] Leo Spitzer, 'The Problem of Renaissance Latin Poetry', in his *Romanische Literaturstudien, 1936–56* (Tübingen, 1955), 442–3.

[93] See above, n. 64.

[94] For its possible implications on reading habits see T. Cave, 'The Mimesis of Reading in the Renaissance', in *Mimesis: From Mirror to Method*, ed. L. O. Lyons and S. G. Nichols Jr. (Hanover and London, 1982), especially 155–6.

preoccupied with *aemulatio* as cultural exegesis in a manner that sets them apart from even the most ingenious twelfth-century imitators of Ovid or the most sophisticated Renaissance followers of Catullus. And both of them set out to show, within the compass of an 'Alexandrian' lyric genre, how a 'modern' poet, by the bold appropriation of his ancient models, could himself attain the status of an alternative classic. 'Alternative', because erotic poetry provided them with an opportunity to voice their revolt against convention, both moral and interpretative. How and why, are plain to see. The erotic mode, superficially so personal, acquires an almost symbolic significance in the hands of Peter of Blois and Johannes Secundus, offering them a uniquely effective means of mediating their passionate engagement with the past.

The history of these creative subversives, from the twelfth century to the Renaissance and beyond, is a history of paradox, a discontinuous tradition of rivalry and antagonism that has little to do with 'edle Einfalt und stille Grösse'. Freedom within classicism was won neither by conformity to the prescriptions of rhetoric nor by anticipation of romantic values. True liberty always was achieved more subtly. In the twelfth, as in the sixteenth century, Athens had to be regained from Alexandria.

Appendix

Ars versificatoria 2. 57 (Munari, 84–6)

Vel si deliciosus erit auditor, dicens quod in multiloquio pretium non est, membrorum descriptionem sic comprehendat:

Respondent ebori dentes, frons libera lacti,
  Colla nivi, stellis lumina, labra rosis.
Artatur laterum descensus ad ilia, donec
  Surgat ventriculo luxuriante tumor.
Proxima festivat loca cella pudoris, amica
  Natura, Veneris deliciosa domus.
Que latet in regno Veneris dulcedo saporis,
  Judex contactus esse propheta potest.
Pes brevis, articuli directi, carnea crura,
  Nec vacua fluitat pelle polita manus.
Ne titulo careat species, sua dona maritant
  Corporeae dotes, effigiale bonum.

Materiae pretium, formae praestantia quaeque
  Membra relativa sedulitate beant.
Materiam picturat opus praedulce, venusto
  Materiae pretio materiata placent.
Non floris pretium marcescit turbine fastus
  Ceu parcat tenerae spina miserta rosae.
Haec facit ad Venerem, mihi tales eligo, tales
  Describit quales Vindocinensis amat.
Hoc pretio Frigios lesit Ledea, rapina
  Priamide, Troje flamma, ruina ducum.
Cur hanc Priamides rapuit si Graecia quaerit,
  Illic Ypolitum pone, Priapus erit.

Johannes, *Basium* 12

Quid vultus removetis hinc pudicos,
Matronaeque, puellulaeque castae?
Non hic furta deûm iocosa canto,
Monstrosasve libidinum figuras.
Nulla hïc carmina mentulata, nulla
Quae non discipulos ad integellos
Hirsutus legat in schola magister.
Inermeis cano basiationes,
Castus Aonij chori sacerdos.
Sed vultus adhibent modo huc protervos
Matronaeque, puellulaeque cunctae,
Ignari quia forte mentulatum
Verbum diximus, evolante voce:
Ite hinc, ite procul, molesta turba,
Matronaeque, puellulaeque turpes!
Quanto castior est Neaera nostra,
Quae certe sine mentula libellum
Mavult, quam sine mentula poetam?

Catullus 16. 1–6 (cf. Martial 3. 69)

Pedicabo ego vos et irrumabo,
Aureli pathice et cinaede Furi,
qui me ex versiculis meis putastis,
quod sunt molliculi, parum pudicum.
nam castum esse decet pium poetam
ipsum, versiculos nihil necessest.

Johannes, *Ad grammaticos, quur scribat lascivius*

Carmina quur spargam cunctis lasciva libellis.
  Quaeritis? *insulsos arceo grammaticos*!

Fortia magnanimi canerem si Caesaris arma,
Factave divorum relligiosa virûm,
Quot miser exceperemque notas, patererve lituras.
Quot fierem teneris supplicium pueris?
At nunc uda mihi dictent quum basia carmen.
Pruriat & versu mentula multa meo!
Me legat innuptae iuvenis placiturus amicae,
Et placitura novo blanda puella viro,
Et quemcumque iuvat lepidorum de grege vatum
Otia festivis ludere deliciis!
Lusibus at laetis procul hinc absistite, saevi
Grammatici, injustas & cohibete manus,
Ne puer, ob molleis caesus lachrymansque lepores,
Duram forte meis ossibus optet humum!

Johannes, *Basium* 1

Cum Venus Ascanium super alta Cythere tulisset,
Sopitum teneris imposuit violis,
Albarum nimbos circumfuditque rosarum
Et totum liquido sparsit odore locum.
Mox veteres animo revocauit Adonidis igneis,
Notus & irrepsit ima per ossa calor.
O, quoties voluit circundare colla nepotis?
O, quoties dixit, 'Talis Adonis erat!'?
Sed, placidam Pueri metuens turbare quietem,
Fixit vicinis basia mille rosis.
Ecce calent illae, cupidaeque per ora Diones
Aura, susurranti flamine, lenta subit.
Quotquot rosas tetigit, tot basia nata repente
Gaudia reddebant multiplicata Deae.
At Cytherea, natans niveis per nubila cygnis,
Ingentis terrae coepit obire globum.
Triptolemique modo, foecundis oscula glebis
Sparsit, & ignotos ter dedit ore sonos.
Inde seges felix nata est mortalibus aegris:
Inde medela meis unica nata malis.
Salvete aeternum, miserae moderamina flammae,
Humida de gelidis basia nata rosis!
En ego sum, vestri quo vate canentur honores,
Nota Medusaei dum iuga montis erunt,
Et memor Aeneadûm stirpisque disertus amatae,
Mollia Romulidûm verba loquetur Amor!

Rudolph Agricola, *Aphthonius' Progymnasmata*

Si quisquam est qui rosae decorem miratur, is animo mihi Veneris reputet vulnus. Amabat enim dea Adonidem, contra Mars ipse deam amabat, parique dea Adonidis amore, quo Mars Veneris tenebatur. Flagrabat deae desiderio deus, dea sequebatur hominem, similis utrorumque ardor tametsi natura inter se ut dii homoque different. Invidens Mars Adonidi perimere ipsum decrevit, unicum abolendi amoris remedium, mortem ipsius arbitratus. Percusso demum Adonide, quum rem dea accepisset, ut succurreret, festinabat, raptoque properanter inter rosaria cursu, spinis incidit, calcemque pedis perstrinxit, defluens autem ex vulnere sanguis, rosae colorem in novam faciem mutavit. Sicque rosa, quae initio fuerit alba, in hunc quo nunc aspicitur transit colorem.

Johannes, *Basium* 16

Latonae niveo sidere blandior
Et stella Veneris pulchrior aurea,
  Da mi basia centum!
  Da tot basia, quot dedit

Vati multivolo Lesbia, quot tulit;
Quot blandae Veneres quotque Cupidines
  Et labella pererrant
  Et genas roseas tuas;

Quot vitas oculis quotque neces geris,
Quot spes quotque metus quotque perennibus
  Mixta gaudia curis
  Et suspiria amantium.

Da, quam multa meo spicula pectori
Insevit volucris dira manus dei
  Et quam multa pharetra
  Conservavit in aurea.

Adde et blanditias verbaque publica
Et cum suavicrepis murmura sibilis,
  Risu non sine grato,
  Gratis non sine morsibus,

Quales Chaoniae garrula motibus
Alternant tremulis rostra columbulae,
  Cum se dura remittit
  Primis bruma Favoniis.

Incumbensque meis mentis inops genis
Huc, illuc oculos volve natatiles
  Exsanguemque lacertis
  Dic te sustineam meis.

Stringam nexilibus tete ego brachiis,
Frigentem calido pectore comprimam
  Et vitam tibi longi
  Reddam afflamine basii.

Donec succiduum me quoque spiritus
Istis roscidulis linquet in osculis,
  Labentemque lacertis
  Dicam conlige me tuis.

Stringes nexilibus me, mea, bracchiis,
Mulcebis tepido pectore frigidum
  Et vitam mihi longi af-
  Flabus rore suavii.

Sic aevi, mea lux, tempora floridi
Carpamus simul; en iam miserabiles
  Curas aegra senectus
  Et morbos trahet et necem.

With l. 5, cf. Catullus 68. 128. With 9 ff. cf. Catullus 68. 17–18. With 17 ff. cf. Ovid, *Amores* 3. 7 (6). 9 ff.

9 The Origin and Development of the Catullan Style in Neo-Latin Poetry

WALTHER LUDWIG

In the year 1594 Friedrich Taubmann, at that time *baccalaureus artium* of the University of Wittenberg and *poeta laureatus*, wrote an *Epithalamion* on the marriage of Prince Johann Sigismund of Brandenburg with Princess Anna of Prussia. It consists of 143 Catullan hendecasyllables. At the beginning the poet answers the objection that a prince should be celebrated in the heroic hexameter by defending the *mollis* (*versus*) *Phaleucus* as the appropriate metre for the princely *nuptiae* (1. 13 ff.):[1]

Cum mollissima sit Venus deorum,
Non versus amet illa mollicellos?
Cum blandissima diva sit deorum,
Non versus amet illa blandicellos?
Aut his est reperire molliores?
Aut his est reperire blandiores?
Aut pro conditione belliores?
Aut ad Cypridis orsa lectiores?
Hoc, pol, hendecasyllabo Phaleuco
Nullus mollior esse blandiorve,
Nullus bellior esse lectiorve
Docti iudicio potest Catulli.

Since Venus is the gentlest of the gods,
Would she not love those gentle little verses?

This paper remains in the form in which it was presented, with the addition of a few footnotes. A more extensive and detailed treatment of the same subject will appear in German under the title '*Catullus renatus*—Anfänge und frühe Entwicklung des catullischen Stils in der neulateinischen Dichtung' as a chapter of my forthcoming book: *Litterae Neolatinae: Schriften zur neulateinischen Literatur*, ed. L. Braun, W.-W. Ehlers, P. G. Schmidt, B. Seidensticker (Humanistische Bibliothek, i. 35; Munich), 162–94.

[1] F. Taubmann, *Melodaesia sive epulum musaeum* . . . (Leipzig, 1604 (1st edn. 1597)), 224. Cf. *Allgemeine Deutsche Biographie*, xxxvii. 433 f., s.v. Taubmann, Friedrich.

Since she is the most charming of the gods,
Would she not love those charming little verses?
Can anything gentler than them be found?
Can anything more charming than them be found?
Can anything more lovely in form than them be found?
Can anything choicer for Venus' lips than them be found?
Lord, nothing can be gentler,
more charming, more beautiful, or choicer
than the Phaleucian hendecasyllable,
according to the judgement of the learned Catullus.

In these verses Taubmann expresses the character of the Catullan hendecasyllable as he saw it.

An analogous description of the characteristics of the *carmen Phaleucium* was given by professors of the University of Giessen in their introduction to poetics, entitled *Poetica* and printed for the first time in 1608.[2] After quoting Scaliger's statement '(Phaleuci) lascivi . . . maxime omnium sunt',[3] they make the following recommendations for the composition of good hendecasyllables. There should be a caesura after the penthemimer, as in *Cui dono lepidum*, or failing that a word break after the second dactyl, as in *Corneli tibi*. Use of *epanodos*, that is, the repetition of the same words, will give a *mira venustas* to the lines. Pleasant effects can be achieved by a return of the initial verse in the form of an *epanalepsis* at the end of the poem, especially if it is combined with another meaning; by similar sounds in successive lines; and by agreeable polysyllables like *elegantiarum*. Taubmann's *Phaleuci* with their diminutives and comparatives are named as a model.

Around 1600 it was expected that a poem in Catullan hendecasyllables should embody certain typical features of Catullan metre, Catullus' distinctive vocabulary, Catullan word-forms and modes of expression, and also certain topics and themes. That there is a characteristic Catullan style in neo-Latin poetry is probably generally known, and I only wish to illustrate it here by two examples which are less familiar. Poems of this kind were written for many occasions and on various subjects. Pontano, Janus Secundus, Lotichius, and others come to mind. But we do not know how and when this neo-Catullan poetry began, how

[2] (Chr. Helwig-C. Bachmann), *Poetica, praeceptis, commentariis, observationibus, exemplis, ex veteribus et recentibus poetis studiose conscripta per Academiae Gissenae nonnullos Professores*, (Giessen, 1617 (1st edn. 1608)), 245 ff.

[3] J. C. Scaliger, *Poetices libri septem* (Heidelberg, 1607 (1st edn. 1561)), 179 (*ib.* 2, cap. 38).

it developed, how widespread it was, and how it influenced vernacular literatures.

There are a few more or less detailed studies of specific topics, such as Catullan poetry in France,[4] or in England,[5] or the general relationship of Catullus and Pontano,[6] but on the whole these questions remain unanswered in recent literature. This is even the case in regard to the beginnings of neo-Latin poetry in Catullan style. The relevant literature tells us that it began in the Italian Quattrocento (some say in the Trecento), and some poets are usually mentioned as having allegedly imitated Catullus. These are always listed, however, without any order or differentiation and without any attempt to determine the origin of this poetic style more closely. In most cases poets like Poliziano, Pontano, Marullo, Beccadelli, Navagero, and Flaminio appear in random lists of such imitators side by side; sometimes Lovato and Mussato are included; and the term imitator is often used so broadly and vaguely that it comprises anything from the use of a Catullan word to variations on whole poems and the choice of certain stylistic features or topics.[7]

The two most recent comprehensive monographs on Catullus—by Ernst August Schmidt and John Ferguson—both have chapters on Catullus' influence, but reveal no progress in this regard. Ferguson writes 'much work needs to be done on the influence of Catullus on the early Renaissance in Italy and France', but does not so much as mention neo-Latin poetry among the fields in which Catullus had an impact.[8] Such reticence is hardly justified, even in view of the less than abundant literature on the subject. It can only be explained by the regrettably persistent blindness of some classicists when confronted with the phenomenon of neo-Latin. Schmidt parades the usual throng of neo-Latin imitators

[4] M. Morrison, 'Catullus in the Neo-Latin poetry of France before 1550', *Bibliothèque d'Humanisme et Renaissance*, 17 (1955), 365 ff.; id., 'Catullus and the Poetry of the Renaissance in France', ibid. 25 (1963) 25 ff.

[5] See the literature referred to by J. Ferguson, *Catullus* (*Greece and Rome* New Surveys in the Classics, 20; Oxford, 1988), 46, 48 f.

[6] A. Sainati, *La lirica latina del Rinascimento* (Pisa, 1919), repr. in *Studi di letteratura latina medievale e umanistica raccolti in occasione del suo ottantacinquesimo compleanno* (Padua, 1972). Sainati pays attention to similarities and discrepancies of tone and emotion in the poems of Catullus and Pontano; he is less interested in specific formal borrowings and does not investigate the literary development of the imitation of Catullus in the fifteenth century.

[7] Cf. M. Morrison (n. 4 above, 1955), 365 f.; *Catull: Sämtliche Gedichte, Lateinisch und Deutsch*, introd. and tr. O. Weinreich (Zürich, 1969 (1st edn. 1960)) p. 69; K. P. Harrington, *Catullus and his Influence*, (New York, 1963) 104; A. Sainati (n. 6 above, 1972), 62.

[8] n. 5 above, 44 ff.

of Catullus, and then draws in broad strokes the outlines of Catullus' influence in the modern period. According to him, the creative reception of Catullus prepared for and accompanied the analogous imitation of Horace's *Odes* and was finally overtaken by the imitation of Pindar.[9] This view is erroneous. Before any such generalizations are hazarded, detailed investigations are necessary, and they have not yet been undertaken.

All this leads me to ask: how was Catullus interpreted after the rediscovery of his works? Where and when did poems in Catullan style come to be composed? What were their motifs? What was considered to be Catullan and what was reproduced? What was the purpose of these poems? And finally, what impact did these poems have on the later Catullan poetry of the Renaissance, neo-Latin and vernacular?

The still lively controversy, whether the Catullus manuscript which is the source of our tradition was brought from France to Verona around 1300 or had always lain in Verona and was rediscovered at about that time, is not immediately relevant to these problems. Suffice it to say that there is no evidence that Lovato knew Catullus as early as 1265, as some scholars believe, under the influence of an article published by Guido Billanovich in 1958. Billanovich thought that the phrases *acrior ignis* and *imbre madet*, which occur in a poetical letter of Lovato from 1268, are proof that Lovato had read Catullus by that time. As I have demonstrated elsewhere,[10] however, these combinations of words go back, not to Catullus, but to Seneca and Statius. But did Lovato and/or Mussato imitate Catullus at a later date? The first problem here is whether certain words and combinations of words, which appear in identical or similar forms both in texts of these early humanists and in Catullus, but also in other ancient authors, derive from Catullus or not. In each case, a close comparison of the different contexts is necessary. Sometimes we shall only be able to decide such questions with varying degrees of probability. But however such parallels are evaluated, the use of a few single Catullan words or phrases in the hexameters and distichs of Lovato and Mussato cannot be interpreted as an attempt to compose poems in Catullan style: it would merely be the use of ancient words, and perhaps thoughts, taken from Catullus as well as from other ancient poets. In 1919 Augusto

[9] *Catull* (Heidelberg, 1985), 16 ff.

[10] 'Kannte Lovato Catull?', *Rheinisches Museum für Philologie*, NF 129 (1986), 329 ff., repr. in Ludwig, *Litterae*, 37 ff.

Sainati wrote that Mussato imitated the *passer*-poem of Catullus 'in una svelta poesiola'.[11] But in that epigram[12] Mussato says only that he does not write poems like Virgil's first *Eclogue* or like Catullus' *passer*-poem, and the phrasing of this statement is not even specifically Catullan, but reminiscent of Ovid, Propertius, Tibullus, and Martial. I recommend that the term 'imitation' be used more restrictively. If we call almost any intertextual relationship an imitation, we shall have to find new terms for those phenomena to which this term more properly applies.

In the second phase we need to consider how far the text of Catullus was known in the fourteenth century, and how it was interpreted. For this purpose I have checked the manuscript tradition, which is well researched,[13] and all the known instances where Catullus is mentioned or quoted in the texts of that century.[14] My conclusion is that the text of Catullus was only slowly diffused and that only a small number of humanists in northern and middle Italy came to know it. They studied this text and sometimes quoted passages from it for moralistic, antiquarian, philological, or poetic reasons, as in the case of Catullus 16 with its distinction between the *vita* and the verses of the poet. The various Catullan metres were observed and defined, some *ioci* appreciated. The evaluation which Petrarch's friend Guglielmo da Pastrengo made is characteristic: 'Catullus, Veronensis poeta, Ciceronis coetanus, librum vario metrorum genere exaratum, multa iocosa et placita continentem scholasticis tradidit.' Petrarch himself once used the term *nugae* in a letter as a formula of modesty for his own writings, perhaps in imitation of the preface of Pliny's *Natural History*. Some scholars have supposed that he used a handful of Catullan motifs in his *Canzoniere*, but the evidence is sparse and not undisputed. In the *Trionfo d'amore*, Catullus appears after Ovid, Tibullus, and Propertius, but Petrarch also introduced Alcaeus, Pindar, and Anacreon into the triumphal procession. He esteemed Catullus as an ancient poet, but Catullus was never among his favourite authors. In the whole century, nobody tried to write poems in Catullan style. Catullus was one of the rediscovered ancient authors and, as such, important; but he was not

[11] n. 6 above, 1972, 62.

[12] See L. Schwab, *Catulli Veronensis Liber* (Berlin, 1886), p. xiv.

[13] See the survey of extant manuscripts by D. F. S. Thomson, *Catullus: A Critical Edition* (Chapel Hill, 1978), 44 ff.

[14] See Ludwig, *Litterae*, 166 f. for more details.

anyone's favourite author. His place was in the second rank: his rediscovery had not yet led to his renaissance.

In the first half of the fifteenth century his text circulated much more widely, although manuscripts were still rather rare.[15] The first warnings were given against reading Catullus in school: he was considered to be too obscene.[16] It was at this time and in this context that Latin poetry in the Catullan style began. In modern scholarship Antonio Becadelli, known as Panormita (1394–1471), is often named as a prominent imitator of Catullus. But I have found no indication either in his collection of poems of 1425, the *Hermaphroditus*, or in the letters of humanists which refer to it, that this is true. Panormita mentions Catullus at the head of a list of ancient poets who, as he says, all wrote poems full of *lascivia*, and he defends his own obscene elegiacs with the Catullan distinction between the *castus et pius poeta* and the *versus molliculi et parum pudici*, which had already been quoted with approval by Coluccio Salutati.[17] Catullus and other ancient poets are thus used to protect and legitimize Panormita's audaciously obscene and sexually oriented poetry. But in the poems themselves very few borrowings from Catullus are to be found, restricted to the use of the verbs *pedicare* and *irrumare*, to certain obscene motifs from Catullus 97, and to the topos of the thousand kisses.[18] Panormita never tried to imitate the Catullan style or the characteristically Catullan hendecasyllables. Martial and the *Carmina Priapea* were more prominent as his models. But he is important in having pointed to Catullus as a patron of frivolous poetry and thus preparing the way for the imitation of Catullus by Pontano.

Before Pontano followed this lead, Cristoforo Landino (1424–1504) came to Catullus by another route in Florence. He wrote a collection of poems, entitled *Xandra*, in 1443/4.[19] It contains mainly elegiacs, but also two hendecasyllabic poems and three Sapphic odes. On the whole a cycle of love poems, it draws primarily on Propertius. But Landino also mentions Catullus among his favourite authors, and in order to show himself in

[15] Thomson (n. 13 above), and F. R. Hausmann, 'Datierte Quattrocento-Handschriften lateinischer Dichter . . . und ihre Bedeutung für die Erforschung des italienischen Humanismus', in *Kontinuität und Wandel . . . Franco Munari zum 65. Geburtstag* (Hildesheim, 1986), 598 ff.

[16] R. Sabbadini, *Le scoperte dei codici latini e greci ne' secoli XIV e XV* (Florence, 1905/14; revised repr. E. Garin, Florence, 1967), i. 201, on Ugolino Pisani.

[17] F. C. Forberg, *Antonii Panormitae Hermaphroditus* (Coburg, 1824), 5 ff.

[18] Ibid. 54 (*Herm.* 1. 13), 61 ff. (*Herm.* 1. 17), and 181 f. (*App.* 2).

[19] A. Perosa, *Christophori Landini carmina omnia* (Florence, 1939).

the role of a passionate and unhappy lover, he uses thoughts and expressions from Catullus 8 and 87.[20] Furthermore, he incorporates structural elements of the Sapphic ode, Catullus 11, into a Sapphic ode of his own.[21] These are, to my knowledge, the earliest attempts to use the conceptual content of specific Catullan poems and to imitate the Catullan style and his characteristic hendecasyllables. Pontano, in his second collection of Catullan poetry, the *Liber Parthenopaeus*, alluded to these poems of Landino's and emulated them in his own Catullan style.[22] But he himself seems to have been influenced more by Panormita, and it is not yet clear how much influence the Catullan poems of Landino had in other respects. Their impact on the later development of the Catullan style in neo-Latin poetry seems to have been small. This is partly because Landino very rarely composed hendecasyllabic poems and did not in any case continue to draw on Catullus in his later poems. In addition, his poems were scarcely known after 1500, not being printed until the eighteenth century.

The credit for establishing and propagating the Catullan style in neo-Latin poetry is to be ascribed to Giovanni (Gioviano) Pontano (1429–1503). We may observe three different phases in the development of his Catullan poetry. He went to Naples in 1448, where he met Panormita, who became his close friend. Stimulated by the *Hermaphroditus*, Pontano wrote a collection of poems, entitled *Pruritus sive de Lascivia*, in 1449, the second version of which he dedicated to Tito (Vespasiano) Strozzi at Ferrara in 1451.[23] The problem with this collection is that neither of its two versions is transmitted in full. But we have various manuscripts which contain different but overlapping sections of it, and some of these poems also appear in the *Liber Amorum* 1, printed for the first time in 1505. By comparing the order of the poems in the different manuscripts and by an analysis of their content, I have been able to some extent to reconstruct the *Pruritus*.[24] It contained introductory and closing poems, and at least seven erotic, in part very obscene, poems, three poems addressed to friends, and

[20] *Xandra* 6 (= A I 16), 35 (= A I 17).

[21] *Xandra* 50 (= A I 30).

[22] Compare Landino, *Xandra* 1 (= A I 13) with Pontano, *Am.* 1. 1 and 28, *Xandra* 6 with *Am.* 1. 9, and *Xandra* 45 with *Am. 1. 4*. The poems of Pontano are quoted according to B. Soldati, *Ioannis Ioviani Pontani carmina* (2 vols., Florence, 1902).

[23] The date of 1449 can be inferred from the epilogue poem *App.* 7. 7 Soldati ('bis denis ... me crescere in annis') in conjunction with his year of birth (1429, not 1426; see S. Monti, 'Il problema dell'anno di nascita di Giovanni Gioviano Pontano', *Atti dell'Accademia Pontaniana Napoli*, NS 12 (1962/63), 225 ff.). Pontano visited Ferrara in 1451: W. Ludwig, *Die Borsias des Tito Strozzi* (Münich, 1977), 33 f., 300 f.

[24] *Litterae*, 173 f.

one invective against a contemporary poet.[25] Ten poems are in Catullan hendecasyllables, and one is a Sapphic ode;[26] glyconics, distichs, and hexameters also occur. The metres and themes indicate that poetry in the Catullan style was intended. The title *Pruritus* and the first poem declare that the purpose of the *novus libellus*, as Pontano calls it, is to make the reader prurient. Catullus and Martial occasionally said that their *versus molles et iocosi* should provoke a *prurire* on the part of the reader.[27] Pontano surprises his readers by proclaiming this aim at the very beginning. His collection stands in the tradition of the *Hermaphroditus*. In both cases the *Carmina Priapea* provided ideas and expressions. But in the *Pruritus* there are also numerous conscious imitations of Catullus. Analysis of its hendecasyllables shows that Pontano observed the stylistic peculiarities of Catullus closely and reproduced them in concentrated and sometimes exaggerated form, employing typically Catullan words of commendation and vituperation such as *elegans*, *venustus*, *illepidus*, *lepos*, *elegantiae*, *facetiae*, *ineptiae*, *nequitiae*; erotic and scatological language; diminutives such as *ocelli*, *labella*, *papillae* and *papillulae*, *medulla*, *puellula*, *aureolus*, *turpiculus*, *cinaedulus*, *solaciolum*; polysyllables such as *basiationes*, *osculationes*, *cacationes*, *mentulatiores fututione*, *elegantiarum*; comparative expressions such as *mundior elegantiorve*, *plus oculis*, *merdosius omnibus latrinis*; generalizing pronouns; particles, intensifications, repetitions, and reversals. In addition, several works are constructed as variations on specific Catullan poems. One, for instance—*Amores* 1. 5 in the printed editions—in which the author plans to give his *nivea columba* to his *puella* is clearly a variation of the *passer*-poems. Sexual connotations in the envisaged or imagined activities of the *columba* indicate that Pontano wanted the poem to be allegorically understood, and that he interpreted the *passer*-poem itself in a manner similar to that of Poliziano some time later. In another poem, *Ad Thaliam* (*Amores* 1. 26), Pontano employs the Catullan motif of *basia mille* in a very special way: he asks the Muse Thalia to give many hundred kisses to a friend of his.

[25] *App.* 9, 7, and 8, *Am.* 1. 5, 7, 25 (?), 29, 30, 31, 33, *App.* 5, *Am.* 1. 26, 27, 34, and 32.

[26] The latter is the *Hymnus ad Noctem* (*Am.* 1. 7); for interpretation see W. Ludwig, 'Humanistische Gedichte als Schullektüre: Interpretationen zu Sannazaro, Flaminio und Pontano', *Der Altsprachliche Unterricht*, 29. 1 (1986), 53 ff., esp. 69 ff.; repr. *Litterae*, 250 ff. The hymn contains one of the earliest uses of the ps.-Platonic motive of the exchange of souls by kisses, a motive which humanists first found in Gellius, *NA* 19. 11, and which was later often used in Renaissance poetry; cf. W. Ludwig, 'Platons Kuss und seine Folgen', *Illinois Classical Studies* 14, (1989), 435–47.

[27] Cf. Catullus 16. 9 and Martial 1. 35. 10 f.

(Panormita had already asked the Lady Elegia to give him thousand kisses.) To my knowledge, no ancient poet was ever kissed by the Muses. Perhaps the modern image of the kiss of the Muses had its origin in Pontano's transfer of Lesbia's kisses to Thalia. In these poems of Pontano love is never sentimental. Once he invites his friend Panormita, 'pater omnium leporum', to laugh with him. Pontano followed Catullus in his general themes: erotic pleasures with *puellae*, witty conversation with *amici*, and rude invectives against enemies. He says of the works of a contemporary poet 'non sunt carmina, sed cacationes' and imagines that the Muse has to hold her nose (*Amores* 1. 32). On the whole, he gave his poems a light and gay, joking and pleasurable, sometimes coarse and very obscene tone.

A second phase of Pontano's imitation of Catullus is to be seen in his *Liber Parthenopaeus*, which he dedicated to his Neapolitan friend Lorenzo Bonincontri da San Miniato in 1457.[28] This book comprised the first twenty-eight poems of his *Liber Amorum 1* as subsequently printed from 1505 on. It can be shown that *Amores* 1. 29–34 were not in the original *Liber Parthenopaeus*, but were added later. These are obscene poems which originally belonged to the *Pruritus*. Pontano did include four poems of the *Pruritus* in the original version of the *Liber Parthenopaeus*, but significantly only those which do not contain elements of crude obscenity and invective. For here he generally dispensed with these indecencies. Again he assembled a variety of metres; eight poems are in hendecasyllables. The poems are mostly erotic in content. Motifs of the Roman elegists occur; but Pontano wanted above all to be compared with Catullus. He wanted his verses to be *molles*, *lepidi*, *leves*, *iocosi*. He states in the closing poem (now *Amores* 1. 28), that he tried to follow the *nequitiae*, the *procacitas*, and the *elegantia* of Catullus, and claims that poems of this sort had not been heard in the last thousand years. He also emulated the Catullan poems of Landino,[29] but seems to have intended to come closer to the frank tone and the stylistic characteristics of Catullus. An analysis of the metres, the themes, the style and the content of the poems shows that he tried to imitate Catullus on all these levels. The difference from the *Pruritus* is to be seen first in the quantity of these

[28] The date is approximate. Lorenzo Bonincontri is mentioned at Naples for the first time in 1456; his wife Caecilia, who was living when Pontano dedicated the *Liber Parthenopaeus*, died in 1458; cf. C. Grayson, *Dizionario Biographico degli Italiani* xii (1970), 209 ff. *Am.* I 1 and 28, two programmatic poems for the opening and closing of the book, are dedicated to Bonincontri.

[29] Cf. above, n. 22.

imitations, second in the avoidance of crude obscenities. But he loved erotic scenes. The general atmosphere of the book is cheerful and optimistic. He may show himself in the role of a frustrated *exclusus amator*, or an unhappy love may provoke the idea of death. But in the same poems, or in the next ones, such emotions are replaced by hopes for more pleasurable events—a reconciliation may occur or the lover console himself with another *puella*, who makes up his loss to his full satisfaction. There is never desperation without recovery, never a passionate love for the one and only beloved. It seems that to Pontano these were the characteristics of Catullus.[30].

These Catullan poems of Pontano's provided an impulse to other humanists to try the Catullan style. The markedly increased distribution of the text of Catullus in the second half of the fifteenth century and the fashion for imitation of the ancients could independently have led to imitations of Catullus. But it seems that Pontano's example was always acknowledged. The second most important imitator of Catullus in the fifteenth century, Michele Marullo (1453–1500), is known to have been a younger friend of Pontano at Naples. Not later than 1489, when he came to Florence, he must also have known Landino's poems. He published two *Libri epigrammaton* at Rome in 1489 and two more at Florence in 1497, which again display various metres, including hendecasyllables.[31] Varied lines of Pontano's *Liber Parthenopaeus* occur.[32] Some of Marullo's poems show the typical stylistic elements of Catullus. He also took specific Catullan poems as models, some of them not yet used by Pontano. But he differed from Pontano in that Catullus is neither explicitly one of his

[30] In his younger years Pontano wrote some explanatory notes to Catullus. Pietro Summonte mentions them in the preface to his edition of Pontano's *De immanitate liber* (in *Opera omnia soluta oratione composita* (Venice, 1518), i, fo. 310v): 'Est mihi praeterea in animo, cognito iam ubique Pontano nostro atque ipsius fama extra omne periculum posita, iuvenileis quosdam eius lusus, si Actio nostro ita videbitur, in lucem proferre, quorum quidem suppuduisse hominem illud declarat, quod eius rei nullam, dum vixit, mentionem fecerit. Hi sunt exquisiti quidam commentarioli in Valerium Catullum, cuius illum constat iuvenem studiosissimum fuisse. Quodsi iniuria mortuo fiat edendis iis, quae ille contemnebat quaeque a viri gravitate aliena omnino videri possint, audacia haec nostra legentium utilitate compensetur. Sunt enim talia, ut neminem omnino ea legisse poeniteat.' But Summonte finally refrained from editing these *commentarioli*, perhaps on the advice of the more scrupulous Actius (Sannazaro). Cf. generally J. Haig Gaisser in *Catalogus Translationum et Commentariorum: Medieval and Renaissance Latin Translations and Commentaries*, ed. F. Edward Cranz (forthcoming), on the early commentaries on Catullus.

[31] A. Perosa, *Michaelis Marulli carmina* (Zürich, 1951).

[32] Compare Marullo, *Ep.* 1. 61. 1 f. with Pontano, *Am.* 1. 4. 1 f., and *Ep.* 4. 2. 1 f. with *Am.* 1. 11. 3 f.

poetical models nor the most prominent of them, and love is only one of several of his themes. Above all, there is a decisive difference between the love poems of Marullo and Pontano, which is obscured when both poets are indiscriminately called imitators of Catullus: Marullo praises the beauty of a certain Neaera, assures her of his unchangeable love, complains about her refusals, confesses his torments and accuses Amor of cruelty. But he never describes the happy pleasures of love, not even in his desires. He only once uses the Catullan term *basiationes*—but he presses his hundred kisses on a picture of his beloved girl, which literally goes up in flames under the heat of his passion (*Ep.* 3. 31). In Pontano's *Liber Parthenopaeus* kisses are a dominant motif, and he shows his readers love play up to and after the moment of climax. Marullo is very conscious of this contrast to Pontano. He explains and defends his position in a poem to an unknown 'Quintilianus' (*Ep.* 1. 62)—there are reasons to suppose that this is a pseudonym for his poetical master Pontano. Quintilianus, according to this poem, had censured Marullo's *liber* as nimium castus . . . nimiumque pudicus', but had approved his *ingenium*. In accordance with the often quoted Catullus 16, he apparently required, even from a *poeta castus et pius*, the writing of *versus molliculi et parum pudici*. Marullo, on the contrary, was in favour of *carmina casta*: they please Apollo and the Muses and, as he says, 'vetat ingenuus verba inhonesta pudor'. The example of Catullus does not lead him to compose obscene poems: 'Tu licet Marsumque feras doctumque Catullum / et quoscumque alios Martia Roma legit / non tamen efficies, ut Phrynae scribere malim . . . / Sit procul a nobis obscoena licentia scripti: / ludimus innocuae carmina mentis opus.' And in accordance with his own poetical practice, he lists the permitted themes of love poetry: 'Sit satis auratos crines laudare Neaerae, / sit satis in duram multa queri dominam / et facere iratum saevo convicia Amori, / . . . caetera Thespiadum prohibet chorus.' Imitation of Catullus in the manner of Pontano was not permissible. Landino probably approved.

The Carmelite Baptista Mantuanus (1448–1516) was even more extreme in his opposition to erotic poetry. He condemned all erotic poetry in his elegy *Contra poetas impudice loquentes*, first printed in 1489.[33] There he states (1. 19) 'Vita decet sacros

[33] E. Coccia, *Le edizioni delle opere del Mantovano* (Rome, 1960). I use the edition *Fratris Baptiste Mantuani Carmelite Theologi, Oratoris, ac Poete clarissimi Prima Parthenice, que Mariana*

et pagina casta poetas', and attacks the Catullan distinction between the *vita* and the *carmina* of a poet. God gave the poetical talent to mankind exclusively 'ad sanctos usus'. The famous Carmelite made no distinction between Pontano's and Marullo's imitation of Catullus. He was adamantly opposed to both.

It is remarkable that, just as the end of the eighties of the fifteenth century (and in his own sixties), Pontano returned to his Catullan poetry. He composed two books, titled *Hendecasyllabi*, between about 1489 and 1501.[34] They were the first books of poems in modern times exclusively devoted to this metre, as well as by far the largest collection of such poems, containing 70 poems altogether, with more than 1800 verses. We should note that Pontano wrote these poems at the very moment when Marullo, with whom he remained on friendly terms, had published his opposing views on poetics and just when Mantuanus, with whom he was in correspondence, had raised the principal objections from a Christian position. Pontano did not care. He chose the title for his new poems *sola metri lege* (he knew that the younger Pliny had used the title *Hendecasyllabi* before him)[35] and commended his two books, in their opening poems, to the inspiration of the Catullan Muse. These poems, dedicated to a friend of his own age, Marino Tomicelli (1429–1515), deal with the pleasurable life at the spa of Baiae near Naples, where the most charming *puellae* and his very best *amici* enjoyed each other's company. Some poems are modelled on specific Catullan ones, but close and specific connections of this kind occur less often than in his earlier Catullan collections. The general Catullan style of Pontano, in which certain Catullan peculiarities such as diminutives, sound plays, and repetitions occur more often than in Catullus, is brought to a maturity of its own. The tone is playful and joking, the atmosphere serene, cheerful, and full of laughter. At the same time, Pontano shrinks from no taboo in his erotic images, although he avoids specifically obscene words. Thus he may—in *Hend.* 2. 19—even visualize the masturbation of a man, his friend Pietro Summonte, by a certain Neaera up to the point of ejaculation. This inclusion of sexual actions not only violates the moralistic poetical code of Marullo, it even goes far beyond Catullus. But Pontano always connects Catullan motifs with such

inscribitur . . . opus insigne De mundi calamitatibus . . . aliud eiusdem Contra poetas impudice loquentes sive scribentes opusculum perelegans (Strasbourg, 1518), fos. LL iv ff.

[34] E. Percopo, *Vita di Giovanni Pontano*, ed. M. Manfredi (Naples, 1938), 162 ff.

[35] Cf. Plin. *Ep.* 4. 14.

descriptions; even here he wants to be seen as following Catullus or at least following a trail blazed by Catullus.

The dominant theme of the *Hendecasyllabi sive Baiarum libri II* is *voluptas*. This term—'illa / quae mentes hominum regit, voluptas' (*Hend.* 2. 21)—occurs in the opening poems of both books and later several times. *Voluptas* appears personified as the last word in the last poem (*Hend.* 2. 38), in which Pontano, like Catullus before him, addresses his *hendecasyllabi* and says good-bye to them. They have accompanied him in his youth and in his old age. Now the fun must come to an end. But he asks the future readers of his *nugae* to wish him and his wife *perpetuas choreas* in the Elysian fields and closes with the following wish to his readers: 'Sic vobis in amore nil amarum, / nil insit nisi dulce, sic amando / et noctes pariter diesque agatis, / assistat lateri et comes Voluptas.' *Voluptas* should here be written with a capital V.[36] It was, of course, known to Pontano and his contemporaries as the leading Epicurean term and was interpreted as meaning 'sensual pleasures'.In Lorenzo Valla's dialogue *De vera voluptate*, composed between 1431 and 1449, the poet Maffeo Vegio had defended the Epicurean position. (At first, Valla had planned to give this role to Panormita.)[37] To defend the *voluptas* of Epicurus as the highest aim, was, in the fifteenth century, only possible under the protection of the *poetica licentia*. The poetical example of Catullus and his distinction between the *castus et pius poeta* and his *versus molliculi et parum pudici* made possible and legitimized for Pontano a poetry of Epicureanism, which he separated from his Christian beliefs. Catullus thus had a certain role in the complex process in which some humanists distanced themselves from the moral precepts and doctrines of Christianity.

In this way the outlines of the origin and early development of the new Catullan poetry of the fifteenth century may be drawn. Other humanists, less prominent as Catullan poets, like Poliziano or Sannazaro, might be mentioned; and often their dependence on Pontano can be demonstrated. At the beginning of the sixteenth century Pontano[38] and Marullo were considered to be the best

[36] As in the edition of L. Monti Sabia, *Ioannis Ioviani Pontani Hendecasyllaborum libri* (Naples, 1978), 138.

[37] A. Kent Hieatt and M. di P. Lorch, *Lorenzo Valla: On Pleasure, De voluptate* (New York, 1977), and M. di P. Lorch, *A Defense of Life: Lorenzo Valla's Theory of Pleasure* (Munich, 1985).

[38] Pietro Summonte wrote in the preface to his edition of Pontano's *De prudentia libri* 5 (n. 30 above, fo. 146^v): 'Fuit enim revera Iovianus non horum temporum foetus, sed tamquam revirescentis iam naturae specimen et in re literaria quidam quasi heros.'

Latin poets of modern times. The influential history of Latin poetry published in 1505 by Pietro Crinito under the title *De poetis Latinis*, and often reprinted, propagated this view.[39] A significant indication of Pontano's reputation as a poet is given by a pair of hendecasyllabic epigrams which Ioannes Pierius Valerianus (Giovan Pietro dalle Fosse, 1477–1558) published in Venice in 1505 and again in 1550[40] and which Paolo Giovio reprinted (without headlines) in his famous *Elogia virorum litteris illustrium*.[41]

Pontano's and Marullo's poems were reïssued several times during the sixteenth century and often read by humanists, regardless of the fact that Erasmus accused both authors of paganism. It has yet to be investigated what impact they had on the further development of Catullan poetry in the sixteenth century, and we have not yet seen the full range of poetry written in this style. Preliminary enquiry makes it clear, however, that Catullan poets of the sixteenth century did not simply study and imitate Catullus himself. They were, in their Catullan poetry, also influenced by their Catullan predecessors in the fifteenth century, especially Pontano. The later Catullan style was, as a rule, not as frank and lascivious as Pontano's. Morals in the sixteenth century were stricter. But a light, gay, playful, and melodious style in the hendecasyllabic metre, for which Pontano provided a model, was always considered to be the characteristically Catullan style. The poems of Dutch poets like Johannes Secundus, Janus Duza senior, and Lernutius, or of French poets like Joachim du Bellay and Jean Bonnefons, or of German poets like Paulus Melissus, Friedrich Taubmann, and Paul Fleming provide abundant evidence that Pontano's Catullan style was imitated again and again. In that his vision and use of Catullus was influential, far into the seventeenth century, in determining what humanists saw in Catullus and how they tried to imitate his poetry Pontano seems to have left

[39] W. Ludwig, 'Julius Caesar Scaligers Kanon neulateinischer Dichter', *Antike und Abendland*, 25 (1979), 20 ff., esp. 23; repr. *Litterae*, 220 ff.

[40] *Pierii Valeriani Hexametri Odae et Epigrammata* (Venice, 1550), fos. 121 ff. I edit in the Appendix the epigrams on Pontano from fos. 124 f. with modern punctuation. There is another noteworthy hendecasyllabic poem on fo. 128, entitled *Prolixitatem non incongruam Hendecasyllabis* which begins: 'Pontani Hendecasyllabi notantur, / ut qui nil habeant Catullianum, / ut qui luxurient nimis, licenter / centenis quoque versibus redundent / . . .' Pierius goes on to defend this expansion by various comparisons and concludes in ll. 21 ff.: 'Sic olim hendecasyllabi, Catullus / quos princeps decorat, breves fuerunt. / Postquam cognita vis leporque in illis, / indultum ingenio est, patentiori / decursum spatio et procaciori / lascivitum animo. Quid invidendum, / ut convivae aliquid sapit palato, / id adponere pleniore quadra?'

[41] *Pauli Iovii . . . Elogia virorum litteris illustrium . . .* (Basle, 1577), 96.

a permanent imprint on later neo-Latin Catullan poetry. The quotation from Taubmann with which I began this paper is a case in point.

Appendix

Pontani Tumulus ad Viatorem

Scriptorum venerabilis magister
Pontanus cubat hic, viator. Illum
Commentus bona verba forte adores.
Ne fac! Huic mala quin precare multa,
Omnem ut surripuit poetices vim!
Nam dum tam bene scribit adque amussim
Aequat omnia perpolitque ad unguem,
Quavis arte decoreque absolutus
Secli terriculum fuit sequentis,
Nemo scribere ut audeat nepotum.

Viator

Quin pacem huic, bona quin precare libans
Thus, costum, violas, crocum sepulcro!
Qui dum tam bene scribit adque amussim
Aequat omnia perpolitque ad unguem,
Quae obsoleverat, ars decorque recte
Scribendi auspiciis renidet huius,
Nostra aetas veteres ut aemulari
Et possit simul et simul praeire.

10 Neo-Latin Imitation of the Latin Classics

G. W. PIGMAN III

Most theorists of imitation, from Seneca and Quintilian to the Renaissance, insist that the good imitator must transform his model into something new and, if possible, something better.[1] The most interesting task for the student of imitative literature is to show the how and the why of this transformation, to describe the interaction between text and model. Only too often, in our rush to make sense of intriguing similarities and differences, we do not pause to consider mundane questions such as, 'Is this text really imitating this "model"? Are the similarities simply coincidences? Could the imitation be an unconscious reminiscence of its "model"? A commonplace rather than an imitation of this particular passage?' And often we do not ask the important question, 'Is this text imitating a text which is already an imitation?'[2] Of course, if one can produce an interpretation which plausibly acounts for the interaction between text and model, the interpretation itself becomes a strong argument in favour of taking the

[1] See my 'Versions of Imitation in the Renaissance', *Renaissance Quarterly*, 23 (1980), 1–32.

[2] These questions are not, of course, peculiar to students of antiquity or the Renaissance. Michael Riffaterre, 'Flaubert's Presuppositions', *Diacritics*, 11 (1981), 2–4, argues that Flaubert's allusion to Neptune's famous 'Quos ego' from *Aeneid* 1 is actually a 'quotation of a quotation'. In this instance it is more difficult to know whether Flaubert is alluding to the 'monographie' which Riffaterre cites or using a commonplace characterization of the classroom tyrant. Riffaterre himself seems to want to have it both ways. For other examples of multilayered allusion see Daniel Javitch, 'The Imitation of Imitations in *Orlando Furioso*', *Renaissance Quarterly*, 38 (1985), 215–39, and 'Sixteenth-Century Commentaries on Imitations in the *Orlando Furioso*', *Harvard Library Bulletin*, 34 (1986), 221–50, and John Hollander, *The Figure of Echo: A Mode of Allusion in Milton and After* (Berkeley, 1981), 79–81, 110–11. A particularly vexing problem for the identification of imitations concerns the use of phrase-books and commonplace books as aids to composition; cf. I. D. McFarlane, 'Reflections on Ravisius Textor's *Specimen Epithetorum*', in *Classical Influences on European Culture A.D. 1500–1700: Proceedings of an International Conference Held at King's College, Cambridge, April 1974*, ed. R. R. Bolgar (Cambridge, 1976), 81–90.

imitation as a conscious allusion to a particular passage, but unfortunately this is not always the case. These difficulties obtain in the study of all Renaissance literature, but I shall suggest that identifying and interpreting imitations is much more difficult in neo-Latin than in vernacular literature. After these general considerations, I shall turn to the theory and practice of one of the best but most maligned of the imitative poets, Marco Girolamo Vida.

At first sight the proposition that neo-Latin imitations are more difficult to identify and interpret may well sound absurd, for a computer which had been fed the Roman poets and almost any collection of Renaissance neo-Latin verse could spit out line after Renaissance line with some phrase from an ancient poem. But actually, that is the heart of the problem: what do we do once we have our massive print-out of similar and identical phrases? It is impossible to believe that all of the repetitions are conscious, much less significant, and often a reader, deafened by the roar of the echo chamber, feels incapable of finding a signal in all that noise. I would hazard the guess that a large proportion of the repetitions is due to coincidence and unconscious reminiscence—large enough, in any event, to raise doubts about 'imitations' and 'borrowings'.

Coincidence probably plays a greater role in poetry than prose, for metre often constrains the choice of words and their position in the line. But on the other hand, metre facilitates memorization, and I suspect that many phrases come unconsciously to the neo-Latin poet because they fit the required metrical pattern, somewhat in the manner of the Homeric formulae. I have not done a Parry-like statistical study, but I have the impression, especially in hexameter poetry, that the phrases which are repeated from the ancients appear most frequently at the beginnings and ends of lines.

Unconscious reminiscence may well pose the greatest obstacle to the student of neo-Latin imitation; it could also worry the poet. Petrarch, who makes a principle of not repeating the words of his predecessors without explicitly citing them or making some significant change, is troubled by his unintentional failures to avoid the footsteps of his predecessors. In a famous letter he tells Boccaccio that authors he read not once but a thousand times, like Virgil and Cicero, do not merely reside in his memory; they have lodged in the intimate recesses of his mind so tenaciously that sometimes their phrases come to him without his realizing whose they are or even that they are not original to him.[3] This explanation of unconscious reminiscence is psychologically convincing. Petrarch is describing a process of identification by which he becomes the son to his parental models. In his desire to 'grow very Hee', to use Ben Jonson's phrase, the imitator loses a sense of the boundary between self and other, just as children unconsciously take on the accent, idioms, and mannerisms of their parents. The comparison, commonplace in writings on imitation, is by no means idle.[4] Petrarch notes that he absorbed as a boy the texts which matter most to him, and one can generalize this experience by recalling the kind of education neo-Latin poets received. Anthony Grafton and Lisa Jardine have demonstrated how thoroughly humanist education, once one looks beyond the puffing of inaugural lectures and the like, was a matter of learning to write Latin like the Romans by means of 'regimented notetaking, rote-learning, repetition and imitation'.[5] Since many of the exercises required of students tested their ability to compose extemporaneously, the chances of previously digested material sliping in unawares were rather high. A poet who has been internalized no longer seems an alien possession but an intimate part of oneself.

In another letter to Boccaccio (*Le familiari* 23. 19) Petrarch tells a charming story about one Virgilian phrase which crept into his work. His young protégé, Giovanni Malpaghini, likes to insert the *ipsissima verba* of the ancients into his own poems. When Petrarch offers a parental admonition against such appropriations,

[3] 'Legi semel apud Ennium, apud Plautum, apud Felicem Capellam, apud Apuleium, et legi raptim, propere, nullam nisi ut alienis in finibus moram trahens. Sic pretereunti, multa contigit ut viderem, pauca decerperem, pauciora reponerem, eaque ut communia in aperto et in ipso, ut ita dixerim, memorie vestibulo; ita ut quotiens vel audire illa vel proferre contigerit, non mea esse confestim sciam, nec me fallat cuius sint; que ab alio scilicet, et quod vere sunt, ut aliena possideo. Legi apud Virgilium, apud Flaccum, apud Severinum, apud Tullium, nec semel legi sed milies, nec cucurri sed incubui, et totis ingenii nisibus immoratus sum; mane comedi quod sero digererem, hausi puer quod senior ruminarem. Hec se michi tam familiariter ingessere et non modo memorie sed medullis affixa sunt unumque cum ingenio facta sunt meo, ut etsi per omnem vitam amplius non legantur, ipsa quidem hereant, actis in intima animi parte radicibus, sed interdum obliviscar auctorem, quippe qui longo usu et possessione continua quasi illa prescripserim diuque pro meis habuerim, et turba talim obsessus, nec cuius sint certe nec aliena meminerim. Hoc est ergo quod dicebam, notiora magis fallere, que siquando forsan ex more recursantia in memoriam redeunt, accidit ut nonnunquam occupato et in unum aliquid vehementer intento animo non tantum ut propria sed, quod miraberis, ut nova se offerant' (*Le familiari*, ed. Vittorio Rossi and Umberto Bosco (Florence, 1933–42), 22. 2. 11–14).

[4] For references, see my 'Versions' (n. 1 above), 9 n. 16.

[5] *From Humanism to the Humanities: Education and the Liberal Arts in Fifteenth- and Sixteenth-Century Europe* (Cambridge, Mass., 1986), 22.

Malpaghini, who had memorized the entire *Bucolicum Carmen* in eleven days, replies that he is following Petrarch's own example. Lines in both the *Bucolicum Carmen* (6. 193) and the *Aeneid* (6. 607) end 'atque intonat ore'. One might pass over this story with a smile or perhaps nod sagely as Petrarch moralizes about the fallibility of human intentions, but what makes the story interesting—and somewhat disturbing—is that this phrase seems to be a clever allusion of the sort dear to interpreters of imitative literature, and yet Petrarch assures us that it is a slip.

Micio, who represents the hated Clement VI, uses the phrase while complaining that Christ is a harsh master: 'ac tolerare famem somnumque sitimque / Edocet, adiungitque minas atque intonat ore.'[6] A few lines earlier Micio had accused Christ of avarice, which connects this passage with its Virgilian original, which Servius takes to describe the punishment of Tantalus for avarice. Tantalus, like the servants of Christ, has to learn to tolerate thirst and hunger because of the threatening presence of the greatest of the Furies. Avarice, thirst, and hunger link the two passages, and these connections suggest a further—Micio belongs in hell, as Pamphilus, who represents St Peter, has already said. Turning a line depicting the torment of Tantalus in hell into a blasphemy which should assure a place in hell for a pope whom Petrarch wishes there is just the sort of clever allusion that appeals to neo-Latin poets and readers of neo-Latin poetry. This coincidence, as I take it, should alert us to the pitfalls of taking a plausible interpretation of the relationship between text and model as evidence that a text is alluding to a model. This passage also shows how difficult it is to avoid the words of precedessors, even for a poet as determined as Petrarch. A line just a few before the one in question ends 'vocibus implet'. This phrase seems as close to 'vocibus implent', the end of a line from the *Aeneid* (11. 274), as a phrase which Petrarch asks Boccaccio to change because it is too similar to the end of another line from the *Aeneid*.[7]

Vida could hardly differ more from Petrarch in his playful attitude towards imitation, his joking about unconscious reminiscence, and his celebration of his thefts from the ancients. Although immensely popular in the sixteenth century and later, he has

[6] *Petrarch's* Bucolicum Carmen, tr. Thomas G. Bergin (New Haven, 1974), 94.

[7] 'Quod hodiernum erat, est decima pastorii carminis egloga, cuius quadam in parte ita scripseram: "solio sublimis acerno" [10. 288]; postmodum vero dum relegeretur, attendi simile nimis esse virgiliano carmini; ille enim ait in septimo divini operis: "solioque invitat acerno" [8. 178]. Mutabis ergo et loco illius pones ita: "e sede verendus acerna" ' (*Le familiari* 22. 2. 22).

never lacked critics, particularly those who take a dim view of his thefts. Even Mario di Cesare, who has done more than anyone to revive interest in Vida, calls *De arte poetica* 'a very commonplace piece of work', and Thomas Greene dismisses Vida in a few sentences, branding his conception of imitation 'mechanical and ahistorical'.[8] Since it seems to me that Greene, the foremost authority on Renaissance imitation, does less than justice to Vida, I should like to re-examine Vida's doctrine and practice of imitation.[9]

Few will be surprised to hear that the example of Virgil is more important to Vida than any body of precepts.[10] It is not, however, sufficiently recognized that Virgil's own imitative practice is the main model for Vida's ideas about imitation, or that Vida regards Virgil as the greatest of poets largely because he sees him as the greatest imitative poet, the poet who constantly surpasses his models. In a letter to the civic fathers of Cremona Vida says that every time he reads the *Aeneid* he thinks that Virgil intended to provide precepts for the art of poetry and to show that the Romans always improved what they received from the Greeks.[11] In *De arte poetica* itself he recommends plundering the Greeks for subject-matter and believes such plundering just as praiseworthy as hitting upon a subject no one has treated before. Why? Because that is what Virgil did (2. 541–57). The recommendation to search for material even in the rough, unpolished poets, illustrated

[8] Mario A. Di Cesare, *Vida's* Christiad *and Vergilian Epic* (New York, 1964), 40; Thomas M. Greene, *The Light in Troy: Imitation and Discovery in Renaissance Poetry* (New Haven, 1982), 177.

[9] 'Unhistorical' is a particularly misleading label, for Vida's recommendation of models for diction is based on his sense of the history of the Latin language and Roman literature. Vida does not try to restrict imitation to a single model, not even Virgil: 'Haud tamen interea reliquum explorare labores / Abstiteris vatum moneo' (3. 193–4). His own practice is eclectic, even though Virgil is the foremost model. And he recognizes that changed conditions require new vocabulary: 'Non tamen omnia te priscis fas fidere, qui non / Omnia sufficient. quaerenti pauca labore / Attentanda tuo nondum ulli audita supersunt. / Nos etiam quaedam iccirco nova condere nulla / Relligio vetat, indictasque effundere voces' (3. 264–8). All quotations are from the first edition, *Marci Hieronymi Vidae Cremonensis De arte poetica lib. III* (Rome, 1527).

[10] Precept and Virgilian practice often went hand in hand. Jürgen Blänsdorf, 'Das Neue in der Kunsttheorie Gerolamo Vidas', in *Stimmen der Romania: Festschrift für W. Theodor Elwert zum 70. Geburtstag*, ed. Gerhard Schmidt and Manfred Tietz (Wiesbaden, 1980), notes that Landino's commentary on Horace illustrates Horatian precept with Virgilian examples (pp. 99–100).

[11] 'Hujus ego quoties, Patres optimi, divinum Poema lego, videtur mihi vir ille non tantum historiam scribere, sed de ea ipsa Arte praecepta tradere voluisse, ut ostenderet, quantum Graecis ipsis, a quibus hanc Disciplinam accepimus, defuisset, quodve de hac Arte est, quod de ceteribus omnibus dici posset, nostros videlicet quae ab aliis accepere, ea semper meliora reddidisse' (*Poemata Omnia* (Padua, 1731), ii. v–vi).

by the simile of purifying muddy water in a cistern, appears to be inspired by Virgil's appropriations from his Roman predecessors, especially Ennius.[12] In the main section on imitation, as I shall show in a moment, Vida makes it clear that Virgil is the master of imitation even though he does not name him. The poem concludes with a hymn to Virgil, who is praised for showing Italian poets how to win trophies from the Greeks.

Furthermore, Vida's transformation of *furtum* from a term of reproach to a badge of honour probably stems from his admiration of Virgil's appropriations from his predecessors. Vida may well have been pleased with the notoriety of his method of imitation because it would have given him one more reason to identify with Virgil.[13] The charges of theft are just what a worshipper of Virgil would exult in; Vida, like Malpaghini following father Petrarch, is showing himself to be a true son. For the most persistent criticism of the *obtrectatores Vergili* was that Virgil plagiarized earlier writers. Donatus' life quotes from Asconius Pedianus Virgil's own rebuttal of those who accused him of stealing from Homer: 'cur non illi quoque eadem furta temptarent? verum intellecturos facilius esse Herculi clavam quam Homero versum subripere' (46). The standard defence of Virgil, such as one finds in Macrobius (*Saturnalia* 6. 1. 6), is the one Vida recurs to: what Virgil took sounds better in him than in his source. He did not steal so much as win the battle for pre-eminence, and to the victor go the spoils.

Although Vida recommends taking subject-matter from the Greeks, he does not go into details, and his recommendation to imitate the *dispositio* of the ancients is even more cursory. The primary discussion of imitation occurs in the context of decorum of diction: the ancients provide a guide to the selection of words appropriate to different kinds of poems. Even while he is giving explicit advice on the use of imitation for the choice of appro-

[12] See the passage beginning 'Nec dubitem versus hirsuti saepe poetae / suspensus lustrare, et vestigare legendo' (3. 196–7). 'Hirsutus' is applied to Ennius by both Propertius (4. 1. 61) and Ovid (*Tristia* 2. 259), and the purification simile seems a more polite version of a remark attributed to Virgil: 'Vergilius dum Ennium legeret a quodam quid ageret inquisitus respondit: "aurum in stercore quaero"' (Cassiodorus, *Institutiones* 1. 1. 8). In any event, Vida is acutely aware of Virgil's borrowings from Ennius, as will be seen shortly, and borrows from Ennius himself. For example, in the *Christias* (2. 316–17), Vida's address to the Muses takes a phrase from the opening of the *Annals* (fr. 1 Skutsch).

[13] Early evidence for censure of the thefts is alluded to by Lilio Gregorio Giraldi: 'Admirari ego soleo . . . M. Hieronymum Vidam . . . quem unum hac tempestate meo iudicio eò pervenisse videtur quò sine Graecis, duce cum primis Vergilio pervenire poeta potest, adeo ut à malevolis, surripere, nedum sumere, dicatur' (*Dialogi duo de Poetis nostrorum temporum* (Florence, 1555), 35). The opposition between *sumere* and *surripere* recalls Cicero's comment on the relation between Ennius and Naevius (*Brutus* 76).

priate words, Vida's own imitative practice is showing how imitations function as allusions. The passage begins with a resounding Virgilian echo, 'Ergo alte vestiga oculis, aciemque voluta / Verborum silva in magna' (3. 179–80); these lines recall the Sibyl ordering Aeneas to seek the golden bough in the dark forest, 'ergo alte vestiga oculis et rite repertum / carpe manu' (*Aeneid* 6. 145–6). The implication seems to be that *delectus verborum* is the key to poetic success.

Vida recommends learning the 'ratio fandi' from one poet, if one is pre-eminent, but also suggests hunting for treasure in other poets, even unpolished ones. It would be easy to assume with a recent translator of the poem that 'longe si quis supereminet omnes' (3. 188) refers to Virgil, but I think it would be a mistake.[14] Vida certainly does believe that Virgil is the greatest of all poets, Greek and Latin, but the context of the present passage is finding words appropriate to the different kinds of poetry. Since Vida says that some words fit some metres better than others and that many words suit the stage but not epic or hymns, he presumably believes that writers of comedy and satire should not take Virgil as their foremost model for diction, although they should doubtless study him for those words which suit all the genres.

Throughout the discussion of imitation, as often in the poem, most famously in the much admired section on imitative harmony, Vida is offering his own writing as an example of his precepts. His imitative practice is eclectic even though Virgil is the foremost model. A particularly interesting example occurs in the hymn to Virgil near the end of the poem and shows how far from mechanical, how clever and complex, Vida's practice of imitation actually is. The youth he has been teaching owes everything to you alone, Virgil, 'Dum tua fida lego vestigia, te sequor unum, / O decus Italiae, lux o clarissima vatum' (3. 584–5). The last line appears, although it is hard to rule out the possibility of unconscious reminiscence, to rework two lines from Virgil. The opening of a speech by Turnus to Camilla provides 'o decus Italiae' (*Aeneid* 11. 508), and the first line of Aeneas' dream speech to Hector the overall pattern of the line and the word 'lux', 'o lux Dardaniae, spes o fidissima Teucrum' (2. 281). Could one imagine a more fitting tribute in a hymn to Virgil which is expressing sole indebtedness to the *princeps poetarum*? But it turns

[14] *The 'De Arte Poetica' of Marco Girolamo Vida*, tr. Ralph G. Williams (New York, 1976), 178.

out that the imitative practice of the lines contradicts the declaration, for they also imitate Lucretius' great hymn to Epicurus, one of the *loci classici* of imitative doctrine, 'te sequor, o Graiae gentis decus, inque tuis nunc / ficta pedum pono pressis vestigia signis' (3. 3–4).[15] Furthermore, the very idea of linking the conclusions of Books 1 and 3 with hymns to the god of poetry and to Virgil probably owes something to Lucretius, who opens his first book with the stupendous hymn to Venus and the third with this hymn to Epicurus. The imitation of the hymn to Venus in Vida's hymn to the god of poetry is so striking that I suspect that Vida, who has just written severely against the dangers which love poses to the young and aspiring poet, wants to reclaim for his spiritualized god of poetry the powers of the all too earthly Venus.[16] And this may also be true of the first line of the *Christias*, which assigns to the Holy Spirit dominion which Lucretius gave to Venus.[17] In any event, Vida is not a single-minded follower of Virgil, and Lucretius figures prominently in his work, as of course he did in Virgil's, which may help to account for Vida's own fondness.

After advising eclectic imitation Vida comes to the most controversial part of his theory and practice—theft from the polished poets, speaking with the very voice of the ancients. He carefully subdivides this section, advocating two very different kinds of theft, one which uses disguise to avoid detection, another which boldly seeks to be caught in the act. He offers two motives for the latter: either to give a dramatically different meaning to the same

[15] Vida may be appropriating other details from the hymn to Epicurus for his hymn to Virgil, although they might be coincidences or commonplaces: calling Virgil 'pater' (Vida 3. 566; Lucretius 3. 9), sipping from his poems ('libant': 3. 572; Lucr. 3. 11), and the declaration that no poet competes with him (Vida 3. 574; Lucr. 3. 5–6). Another phrase from Lucretius' hymn, 'depascimur aurea dicta' (3. 12) recurs in Vida's discussion of imitation (3. 211).

[16] 'Te sine nil nobis laetum, nec amabile quicquam' (Vida 1. 549): 'nec sine te quicquam dias in luminis oras / exoritur neque fit laetum neque amabile quicquam' (Lucr. 1. 22–3).

'Ipsae etiam volucres vario tua numina cantu / Testantur, pecudesque ferae, mutaeque natantes / Ad tua iussa citae properant' (Vida 1. 550–2): 'aeriae primum volucres te, diva, tuumque / significant initum perculsae corda tua vi. / inde ferae pecudes persultant pabula laeta / et rapidos tranant amnis' (Lucr. 1. 12–15)

'Salve hominum dulcis requies, divumque voluptas' (Vida 1. 560): 'Aeneadum genetrix, hominum divumque voluptas' (Lucr. 1. 1), 'Calliope, requies hominum divumque voluptas' (Lucr. 6. 94).

[17] 'Qui mare, qui terras, qui coelum numine comples' / Spiritus alme' (*Christias* 1. 1–2): 'alma Venus, caeli subter labentia signa / quae mare navigerum, quae terras frugiferentis / concelebras' (Lucr. 1. 2–4). Vida may be intimating the superior power of the Holy Spirit, which fills land, sea, and sky, but it is hard to be positive that he has Lucretius in mind because there are other ancient passages which resemble his opening lines (cf. Di Cesare (n. 8 above, 157–60)).

words or to score a victory over the ancients by improving on something poorly composed. Then he exhorts his charges to gird for theft, to snatch booty wherever they find it, and contrasts the premature deaths of works by poets who refuse to steal with the fame which his own thefts will assure him. Immediately after the section on imitation follows the warning that the ancients cannot provide words for all occasions, a tacit acknowledgement of the historical distance from antiquity and firm evidence that Vida is no slavish follower of the ancients.

Once again, the writing itself cleverly illustrates the precepts which are being presented. Let us begin with the passage on disguise (3. 217–22), which, given the difficulties of distinguishing a disguised imitation from a coincidence or unconscious reminiscence, is harder to pin down. In fact, the number of possibly relevant passages from ancient poetry is bewilderingly high.

> Cum vero cultis moliris furta poetis,
> Cautius ingredere, et raptus memor occule versis
> Verborum indiciis, atque ordine falle legentes
> Mutato, nova sit facies, nova prorsus imago.
> Munere (nec longum tempus) vix ipse peracto
> Dicta recognosces veteris mutata poetae.

The primary source, in as much as it makes sense to speak of a primary source, appears to be the attempt in the *Aeneid* (8. 209–12) by the thief Cacus to hide the stolen oxen from Hercules.

> atque hos, ne qua forent pedibus vestigia rectis,
> cauda in speluncam tractos versisque viarum
> indiciis raptos saxo occultabat opaco;
> quaerenti nulla ad speluncam signa ferebant.

Vida is using both methods for disguising thefts, changing the signs ('indicia') of the words, by which he presumably means the forms of the words, and changing their order. The imperfect 'occultabat' becomes the imperative 'occule', and the past participle 'raptos' becomes the fourth declension noun 'raptus'. The phrase 'versis indiciis' remains the same, but each word occupies a different metrical position in its line. So far so good, but 'memor occule' occurs in the same place in a line from the *Georgics* (2. 347), and I have no idea whether Vida is borrowing the phrase or remembering it unconsciously. 'Ordine . . . mutato' is intriguing, for Lucretius uses the phrase or 'permutato ordine' five times (1. 677, 686, 801, 827; 5. 185), always with the words immediately adjacent; Vida splits the noun and its modifier across lines. The phrase is such a natural one in the context, however, that again it is hard to be confident that Vida is deliberately practising what he preaches. And the same is true of 'nova facies', which recalls Aeneas to the Sibyl, 'nova mi facies' (6. 104) and 'nec longum tempus', which occurs in a passage in the *Georgics* (2. 80) that may well be the source for the transplanted fruit simile a few lines farther on. The phrase inserted between 'ordine' and 'mutato' is a different matter. 'Falle legentes' recalls another Virgilian phrase, 'nec miseros fallunt aconita legentis' (*Georgics* 2. 152), and the change of the meaning of 'legentis' from 'gatherers' to 'readers' is probably the sort of trick Vida is recommending. The most tantalizing phrase in the passage is 'veteris . . . poetae'. Is Vida alluding to Horace's lengthy complaint in the epistle to Augustus about the unmerited prejudice the Romans have in favour of the old poets (*Epistles* 2. 1. 34–44)? Or is he alluding to Propertius' tribute to Virgil, 'tu canis Ascraei veteris praecepta poetae' (2. 34b. 77–8)? Or is the use of the phrase coincidental or an unconscious reminiscence? Vida can joke that successful disguise will lead even the disguiser to forget whose work he is plundering, but that is hardly consolation for the reader who wants to track down the imitative poet and who might at this point want to modify Schlegel and cry 'who will save us from all of these echoes'?

Fortunately, the passage on undisguised imitation is not so dizzying. Vida ingeniously and allusively implies that Virgil is the master of emulation, imitation which seeks recognition for victory over its model.[18] Improvement on poorly used material resembles successfully transplanted fruit or the transfer of the Trojan *penates* to Italy (3. 234–42):

sic regna Asiae, Troiaeque penates
Transtulit, auspiciis Phrygius melioribus heros
In Latium, quamvis (nam Divum fata vocabant)
Invitus, Phoenissa, tuo de litore cessit,
Nec conubia laeta, nec incepti Hymenaei
Flexerunt immitem animum. tu victa dolore
Occidis, et curae vix ipsa in morte relinquunt.
Numquam o Dardaniae tetigissent vestra carinae
Littora, fors nulli poteras succumbere culpae.

[18] Fourteen years after the publication of *De arte poetica*, Bartolomeo Ricci explicitly claims that Virgil is the master of emulation and examines his improvements of Ariadne's speech in Catullus 64 to prove his point. See *De Imitatione Libri Tres* (1541; repr. Venice, 1545), 43^{v} ff.

This passage is unmistakeably a pastiche of Virgil. Vida is obviously appropriating lines from the *Aeneid* in the first way he has just recommended, a method which John Sparrow once termed 'literary transubstantiation'.[19] He completely transforms the significance of Virgil's words by using them to illustrate the type of imitation that tries to outdo its model. What is not obvious is that Vida is imitating lines which he knows are already imitations: the three of Virgil's lines which are closest to Vida's imitate Catullus.[20] Macrobius (*Saturnalia* 6. 1. 42) pointed to the dependence of Dido's 'felix, heu nimium felix, si litora tantum / numquam Dardaniae tetigissent nostra carinae' (*Aeneid* 4. 657–8) on Ariadne's exclamation in Catullus, 'Iuppiter omnipotens, utinam ne tempore primo / Cnosia Cecropiae tetigissent litora puppes' (64. 171–2). The other imitations, as far as I know, were not pointed out in Virgilian commentary until after the publication of Vida's poem, but the verbal resemblances are much closer. The second of Vida's clever transpositions comes from the same speech of Ariadne. She tells Theseus he did not promise to desert her, 'sed conubia laeta, sed optatos hymenaeos' (64. 141). Dido beseeches Aeneas 'per conubia nostra, per inceptos hymenaeos' (4. 316). Vida substitutes Catullus' 'laeta' for Virgil's 'nostra': 'nec conubia laeta nec incepti Hymenaei'. The most amusing of Vida's double allusions is 'invitus, Phoenissa, tuo de litore cessit'. This modifies Aeneas' moving protest to Dido in the underworld, 'Invitus, regina, tuo de litore cessi' (6. 460), which in its turn modifies a line from Catullus' 'Coma Berenices', in which the lock offers the excuse, 'Invita, o regina, tuo de vertice cessi' (66. 39). This last example, which scandalizes some classical scholars so much that they prefer to speak of Virgil's unconscious reminiscence, is actually more than a double allusion because Catullus' poem, as he writes in 65, is a translation of a poem by Callimachus. Although only a few fragments of this poem were known during the Renaissance (the original of Catullus' line is still lost), it had attracted considerable attention a generation before Vida because Poliziano used a fragment preserved in the scholia to Apollonius Rhodius to restore 'Chalybon' in Catullus' translation.[21] In *De arte poetica* Vida is thus imitating an imitation

[19] 'Latin Verse of the High Renaissance', in *Italian Renaissance Studies: A Tribute to the late Cecilia M. Ady*, ed. E. F. Jacob (London, 1960), 366.

[20] cf. Antonio La Penna, 'La teoria dell' 'arte allusiva' nel *De arte poetica* di Girolamo Vida', in *Tradizione classica e letteratura umanistica: Per Alessandro Perosa*, ed. Roberto Cardini *et al.* (Rome, 1985), 643–50.

[21] Anthony Grafton, *Joseph Scaliger: A Study in the History of Classical Scholarship* (Oxford,

of a translation of a lost original. Furthermore, the earlier version of *De arte poetica* has an extra line after 'in morte relinquunt': 'semianimesque micant oculi lucemque requirunt'.[22] This turns out to be an unmodified line from Ennius, which is quoted by Servius on *Aeneid* 10. 396, 'semianimesque micant digiti ferrumque retractant', with the remark that Varro Atacinus had used the line without change.

This earlier version of the poem is considerably longer, but Vida did add one passage to the section on imitation, a bold affirmation of his own imitative practice (3. 257–61):

> Saepe mihi placet antiquis alludere dictis,
> Atque aliud longe verbis proferre sub iisdem.
> Nec mea tam sapiens per sese prodita quisquam
> Furta redarguerit, quae mox manifesta probabunt
> Et nati natorùm, et qui nascentur ab illis.

Attacks on Vida's imitations began quite early. Giraldi, in *De poetis nostrorum temporum*, probably written around the same time that Vida was working on the earlier version of his poem, says that the malevolent accuse Vida of plagiarism. Against charges like this Vida takes a defiant stance; he has no intention of hiding his thefts but glories in revealing them. Characteristically, Vida relegates a major part of his self-defence to a quotation from Virgil, apparently the only line in the poem which he appropriates without any modification. 'Et nati natorum et qui nascentur ab illis' comes from the response of Apollo's oracle, the prophecy of Trojan dominion over Italy (*Aeneid* 3. 98). Virgil lifted the line from Poseidon's prophecy in the *Iliad* (20. 307–8) that Aeneas and his descendents would rule over the Trojans. But the imitative chain does not stop with Homer, for Servius remarks, 'sane hic versus Homeri est, quem et ipse de Orpheo sustulit, item Orpheus de oraculo Apollinis Hyperborei. mire autem sequenti versu hoc ostendit dicendo *haec Phoebus*'.[23] It is hard to imagine a better defence of imitative theft than this genealogy, a defence which in its learned allusiveness would doubtless win the admiration, if not of contemporary scholars of imitation, at least of that most Alexandrian of the great poets, Virgil.

1983), 33, 97, 103. In 'Paci', ll. 19–20, Vida imitates the line Poliziano had emended: 'Ut omne telum pereat, ut chalybum genus / Haud nominetur amplius' (*Poemata Omnia*, ii. 151); 'Iuppiter, ut Chalybon omne genus pereat' (Catullus 66. 48).

[22] Williams (n. 14 above), 256.

[23] Macrobius quotes the passages from Homer and Virgil (*Saturnalia* 5. 3. 8), and Erasmus cites them and others in his discussion of the proverb, 'Et Nati Natorum' (*Adagia* 3. 1. 93, in *Opera Omnia* ii. 5 (Amsterdam, 1981), 96).

11 Poetry in a Roman Garden: The *Coryciana*[1]

JOZEF IJSEWIJN

Castalidum pater hortatorque Sororum Corycius.
(M. H. Vida, *Coryciana*, Poem 397. 16–17)

On 25 February 1515 Michael Hummelberg, a South German student of canon law in Rome, sent a letter to the well-known humanist Henricus Bebel promising to forward to him, as soon as they came from the press, 'Poemata in Aram Coritianam a multis nostri aevi poetis hic conscripta' ('Poems about the Corycian altar written here by many poets of our time'). Indeed, he wrote, Corycius (that is to say, Monsignor Johannes Goritz of Luxemburg, a prelate of the Curia and a popular Maecenas in Rome—'father and fosterer of the Muses', in Vida's words) wanted to print them 'propediem' ('immediately'). Bebel was never to see these poems in print, since he died on 31 March 1518 and the *Coryciana* volume was not published until July 1524. By then several poets—Silvius Laurelius, Johannes Maria Cataneus, Janus Vitalis, Caius Silvanus Germanicus, Fabius Vigil, and Blosius Palladius—had tried three or four times to persuade Goritz to release his collection for publication and to allow them to prepare the texts and to see them through the press. The final result, however, was a well-printed volume containing several

[1] Recent literature on the *Coryciana* poems is rather scarce. See Rosenna Alhaique Pettinelli. 'Punti di vista sull'arte nei poeti dei *Coryciana*', *La Rassegna della letteratura italiana*, 90 (1986), 41–54, who cites the older literature and makes excellent remarks on the general cultural background, on certain recurrent themes, and on ancient influences such as that of Pliny the Elder; and J. IJsewijn. 'Puer Tonans: de animo christiano necnon pagano poetarum, qui "Coryciana" (Romae, 1524) conscripserunt', *Academiae Latinitati fovendae Commentarii*, 12 (Rome, 1988), 35–46, on the problem of how pagan and Christian elements coalesce in the poems.

hundreds of Latin poems ranging from two-line epigrams to lengthy hymns.[2]

In contradiction of what Hummelberg had announced in 1515, the editors of the 1524 volume claim that Mgr Goritz had long thwarted their plans to publish the poems and that they had had to steal his manuscripts, as it were, from under his pillow. This must certainly be a literary exaggeration, a veiled commonplace to safeguard Monsignor's modesty against the invidious criticism that was never far away in those curial milieux. Nevertheless one can hardly doubt the fact that Goritz, for some reason, changed his mind between 1515 and 1524. The Corycian story begins in 1512, when the sculptor Andrea Sansovino was commissioned by the Luxemburg prelate to carve a statue of Saint Anne, the Holy Virgin, and the Child Jesus—a very popular group, as we know, in the late mediaeval sculpture of the Rhine-Meuse area in which Goritz had his origins (and not unknown in Italian art either, as Alhaique Pettinelli reminds us). The statue was placed in the church of S. Agostino, near Piazza Navona, at that time recently rebuilt by command of Pope Sixtus IV. Today one can still admire the image in that church. Every year on 26 July, the feast of Saint Anne, Monsignor's friends gathered to celebrate both the sculpture and their patron's piety, in Latin and (now lost) Greek poems. Moreover, the collection provides evidence that occasional visitors to Goritz's mansion also thanked their host with similar poems. Such is the case with Ulrich von Hutten, who had to flee from Rome before the feast—we know that he was already in Bologna on 31 July—but even so has five of his epigrams published in the volume.

I surmise that at the beginning of the *Annalia Coryciana* Mgr Goritz may have fostered the idea of doing something similar to what had been happening since 1509 at the other end of Piazza Navona. There a piece of ancient sculpture, the famous Pasquino or Pasquillus, had been erected next to the palace of Cardinal Oliviero Carafa. At certain moments of the year Latin and, at a second stage of its history, Italian satirical poems were attached to it; and from 1509 onwards the Cardinal had them collected and published in slim annual volumes.[3] Why in the end Goritz

[2] The whole story of the *Coryciana* has been told in detail by José Ruysschaert, 'Les Péripéties inconnues de l'édition des "Coryciana" de 1524', in *Atti del Convegno di studi su Angelo Colocci: Jesi, 13–14 settembre 1969* (Jesi, 1972), 45–66. Hummelberg died in 1527, not in 1519 as stated by Ruysschaert, n. 54.

[3] See Anne Reynolds, 'Cardinal Oliviero Carafa and the Early Cinquecento Tradition of the Feast of Pasquino', *Humanistica Lovaniensia*, 34A (1985), 178–208.

abandoned his original plans we cannot know: perhaps he did not, after further consideration, want to put his pious initiative on the same footing with the rather vulgar and not always very respectable figure of Pasquino. On the other hand we may now be happy that Goritz's friends ultimately succeeded with their editorial plans. Otherwise this exceptional mirror of Roman literary life in the early Cinquecento most certainly would have perished in the Sack of Rome (1527). The Sack also led, as is well known, to the tragic end of old Monsignor Goritz himself, who died during his flight towards the homeland which he had not seen for many decades.

Apart from a final survey of the Corycian poets by Franciscus Arsillus (*De poetis urbanis*, fos. KKiiii^{r}—MMiv^{v}), the collection consists of three hundred and ninety-nine poems in a variety of metres. Dactylic hexameters, elegiac distichs, and hendecasyllables prevail, but lyrical stanzas are not wanting. Although some pieces are quite charming, the overall value of the *Coryciana* volume resides less—in my opinion—in its strictly literary merits than in its indisputable cultural significance, and it is easy enough to see why that is so. Bring together a couple of hundred occasional pieces sent to any person or institution celebrating a happy event or mourning a loss and try to read them all in one sequence. The experience will soon prove to be deadly dull. But if such a collection turns up after several centuries when most other evidence about the person or the institution concerned has disappeared, it may become a first-rate source of information which will allow us to reconstruct an entire political, academic, or literary circle and will teach us much about the ideals and opinions cherished or current at the time. So it is with our *Coryciana*. Anybody who reads the entire volume, from Blosius' dedication letter to Fabius Vigil's final *Potatio*, will have a very hard time. Before long he will begin to detest Sansovino's extraordinary artistic skill, Corycius' unmatched piety, and all the poets' brilliant talents. But these pieces were not written to be read in such a way. They were offered to Goritz in small batches or even as single poems, presented and discussed on various occasions among different groups of guests over a total span of ten or more years. And even so, some of the poets themselves sensed the difficulty of avoiding pointless repetition. That was the sentiment voiced by Aurelius Fossa in Epigram 229, which is a *refutatio* formulated with some elegance (fo. Civ^{r}):

Marmora, templa, arae cum iam sint omnia plena
  Carminibus, tabulis, vatibus et titulis.
Infelix quid agam vates? Num scribere quicquam
  Debeo? Quid scribam? Scribere nescio quid!
Corycii statuae, veniam mihi ferte silenti,
  Aut mihi materia carmina digna date.

Since marbles, churches, altars are all filled with poems, paintings, poets, and inscriptions, what can an unhappy poet such as I do? Ought I to write something? If so, what? I am at a loss! Oh images of Goritz, either forgive my silence, or inspire me to verses worthy of my subject.

The same thought was uppermost in Hutten's mind when he exclaimed (350. 6–7, fo. Xii^{v}):

Da mihi, Musa, aliquid novum, dedisti
Quod non his aliis prius poetis.
'Consumpta est,' ait, 'ut nihil supersit,
Omnis copia tot prius poetis.'

O Muse, grant me something new, something you have not yet given to these other poets. 'All the stock themes have been used up before now', she says, 'by so many poets. There is nothing left.'

Sometimes, it is true, this desperate situation gave a poet new inspiration: Silvanus Germanicus introduced himself as gleaning fallen ears of corn and grapes left by his fellow harvesters. Appropriately, his epigram (280. fo. Riv^{v}) bears the title *Racematio*, 'Grape gleaning':

Pauperies inter culmos a messe relictas
  Sollicita spicas colligit aegra manu;
Pinguis et in praedam vindemia linquit egenis
  Uvas, pampineis quae latuere comis.
Omnia Corycii complerunt horrea laudum
  Ingenio vates uberiore sacri.
Sic ego demessos intrabo novissimus agros.
  Vatibus et claris quae cecidere legam.

The woebegone poor collect with eager hand the ears left by the harvesters among the stubble. The rich vintager leaves, as a booty for the needy, grapes hidden among the foliage of the vines. The hallowed poets, of fuller vein, have filled all Goritz's barns with their praises, so that I shall be the last to enter fields which have been reaped, and shall only glean what has been let fall by more famous authors.

If, then, the poetry-lover will find the collection boring to read, he will find it far less so if he uses it as a key to the world of academic life in Renaissance Rome just before the Sack and to the changing climate resulting from the Reformation and the Counter-Reformation. Few, if any, written documents preserve such a complete and precise picture of the social role and impact of Latin poetry and classical culture on Medicean Rome. We know, of course, that exalted prelates as well as leading bankers and other important people enjoyed, or at least pretended to enjoy, Latin verse. Other works by members of the Corycian circle offer ample evidence of this consorting or flirting with Latin among Roman VIPs. There are Aegidius Gallus' and Blosius Palladius' poems for Agostino Chigi as he built his villa on the banks of the Tiber (now the Villa Farnesina).[4] There are the poems of Silvanus Germanicus on a marble statue of Leo X and on the election of Clement VII,[5] the poetic tears shed at the funeral of a rich Roman youth Celsus Mellinus (d. 20 November 1519),[6] Andreas Fulvius' poem *Saecularis sive iubilaeus annus* of 1525, Janus Vitalis' *Sacrosanctae Romanae Ecclesiae Elogia*.[7]

The *Coryciana* reveal the extent of this infatuation with classical poetry by the sheer number of authors involved: more than a hundred. Among them one finds the most illustrious names of the age: Pietro Bembo, Baldassare Castiglione, Marcus Hieronymus Vida, Angelo Colocci, Paulus Jovius (Giovio), Ulrich von Hutten;[8] and scores of minor ones, both Italian and German. And at this point one must not forget that Goritz was by no means the only patron of writers and scholars in Rome, and probably not even the most important.

For our purpose the *Coryciana* are particularly interesting,

[4] See Mary Quinlan-McGrath, 'Aegidius Gallus, *De Viridario Augustini Chigii Vera libellus*: Introduction, Latin Text and English Translation', *Humanistica Lovaniensia* 38 (1989), 1–99. Blosius' poem will be published by the same editor in the 1990 volume of the same journal.

[5] Today the only known copy is in the Rare Books Department of the University of Pennsylvania Library, which kindly sent me a xerox. The volume bears the printer's date, Rome, 21 December 1524. Part of the first poem has been published and commented upon by H. H. Brummer and T. Janson, 'Art, Literature and Politics: An Episode in the Roman Renaissance', *Konsthistorisk Tidskrift*, 45 (1976), 79–93.

[6] M. Bevilacqua, 'De Celso Mellino eiusque in Longolium oratione', *Latinitas*, 23 (1975), 53–6; L. V. Ryan, 'Milton's *Epitaphium Damonis* and B. Zanchi's Elegy on Baldassare Castiglione', *Humanistica Lovaniensia*, 30 (1981), 108–23.

[7] G. H. Tucker, 'Sur les *Elogia* (1553) de Janus Vitalis et les *Antiquitez de Rome* de Joachim Du Bellay', *Bulletin d'Humanisme et Renaissance*, 47 (1985), 103–12.

[8] On Hutten's contribution see M. Asztalos and T. Janson, 'Hutten correctus: An Example of Humanist Editorial Practice', *Eranos*, 76 (1978), 65–9. The Roman editors corrected several metrical errors and awkward expressions in the original texts left by Hutten in Rome on his hurried departure. The original versions were published in the collection of Hutten's epigrams by Johannes Miller at Augsburg, 2 January 1519, five years before the Roman *Coryciana*.

because they illustrate how Augustan Rome had become a model for its Medicean successor. So much is clear from the very title of the volume, which is derived from the poetic name of Mgr Goritz. He is a new *senex Corycius* in all respects. The sympathetic figure of the old Corycian (i.e. Cilician) gardener living peacefully in the south of Italy is familiar enough to all readers of Virgil's *Georgics*, Book 4. In Blosius' dedicatory letter systematic parallels with Mgr Goritz are established: both are foreigners settled in Italy ('in homine natione externo Romanum ingenium'—fo. Aiii^{v}: 'a Roman temperament in a man of foreign origin'—describes the prelate); both are old ('annis aquaevum'—fo. Aii^{r}: 'equal in years'); both cultivate their gardens (an allusion to Goritz's *vigna*); and both are immortalized by poets.[9] The conclusion, then, is: 'Quodsi P. Vergilius in tua aut tu in illius tempora incidisses, habuisset ille quidem quid de te . . . concineret' (fo. Aiii^{v}: 'If Virgil had lived in your age or you in his, he certainly would have known what to sing about you'). But the Virgilian connection does not stop there. How, indeed did the poems come into Blosius' hands for publication? That we are told again, in Virgilian imagery, at the end of the letter (fo. Bi^{v}):

Librum itaque istum, quem tu capsula occlusum tenebas in tua cellula ad laevam manum, sopito nuper tibi subripui et quasi a Sileno dormiente Vergiliani pueri, sic ego a Corycio sene aeterna carmina clam extorsi.

Recently I have purloined during your sleep that book which you kept locked in a box to your left hand in your room; just as those boys in Virgil did with the sleeping Silenus, so I took away secretly the immortal poems from the old man.

Here we are reminded of an amusing story in Virgil's sixth *Eclogue*: two boys, Chromis and Mnasyllos, having found Silenus asleep in his cave, bind him with own wreaths. By doing so, they hope to oblige him to sing some of his songs for them. And indeed, when he awakes, he readily complies with their wish (ll. 25–6): '"Carmina quae vultis cognoscite" . . . / . . . simul incipit ipse' ('"Hear the songs you desire" . . . and immediately be began').

[9] One poet, Mariangelus Accursius (*Protrepticon*, fo. Biii^{r}–Ci^{v}) goes even further and traces Goritz's origins back to the Corycian cave in Cilicia itself (ll. 70–3): 'Hoc antro Omnipotens certo dedit ordine fati / Corycium prodire patrem, quem collibus altum / Axe sub extremo, divum domus hospita Roma / Gentibus excoluit . . .' ('The Almighty by a fixed disposition of fate made Father Goritz come out of this cave; he was educated in the hills of the North, and hospitable Rome, the home of the saints, distinguished him in the sight of the nations').

The classical Roman interpretation of life and culture in the Goritz circle is a pervasive feature of the *Coryciana*, together with the idea of a Christian renewal of pagan traditions. Parallels are often drawn explicitly. Epigram 63 (fo. Giii^{v}) by Janus Vitalis, one of the major contributors, likens Goritz to Numa, the wise king of Rome, Sansovino to Phidias (obviously, no famous name of a Roman sculptor was available!), and the poets to Virgil. Similarly, Marcus Antonius Casanova (Epigram 162 on fo. L iv^{v}) ascribes to Goritz's piety the same effect as Maecenas' patronage; he too sees a new Phidias, a second Apelles (Raphael, that is, who painted a fresco above the statue), and a return of Virgil's spirit. Other poets are not prepared to go as far as that and, more modestly, pray that ancient Rome's glory may eventually be restored. Such is the final supplication of the beautiful and moving hymn by a Roman contributor, Delius Hieronymus Alexandrinus (Poem 379. fos. BB ii^{r}–iv^{v}):

Si veteres proavi, quanquam caligine mersi,
Adversus Superos semper pietate fuere
Insignes, culmen viguit si maxima semper
Relligionis, et est modo culmen Roma deorum,
Sique pii quicquam, qui te cecinere, poetae
Profecere suis numeris, iam respice, Virgo,
Romanas arces, vasta et Capitolia passim
Obsita nunc dumis silvestribus, aurea quondam
Respice, Diva; iterum rerum moderetur habenas,
Terrarumque orbis solitum sit Roma theatrum!

If our ancient forefathers, though they lived in darkness, have always been known for their piety towards the gods, if mighty Rome has always been a bastion of religion, and still is a stronghold of the saints, if the godly poets who sang your praises have somehow fostered piety with their verses, then, O Virgin, gaze again on the Roman citadel and the vast Capitol—once golden, but now overgrown with wild scrub. Turn back your gaze, Holy Virgin; let Rome govern the world again and let it be the theatre of the earth it used to be!

The divergent approaches to the state of Rome by different poets are interesting enough in themselves. Some of them see the classical poets reborn everywhere, such as, again, Janus Vitalis (poem 8, fo. Ciii^{rv}):

I nunc, crede fuisse saecla avorum
Natura ingenio arte doctiori
Nostris temporibus peritiora!
Nimirum hic tuba maxima Maronis

Inflatur, lyra personat Horatii
Et centum numerantur hic Catulli
Et compti totidem meri Tibulli
Ut dicas merito: 'Revixit aetas,
Aetas pristina, pristini poetae.'

Come now, will you believe that the generations of our fathers owed to a more learned nature, talents, or art greater intelligence than ours? To be sure, Virgil's loftiest style here resounds, here Horace's lyre rings clearly and here Catulluses are counted by hundreds with as many impeccably elegant Tibulluses. Thus you may truly say: 'Antiquity and the ancient poets have been reborn.'

Others are more cautious and, like Blosius in his dedication (fo. Bi^{r}), candidly admit 'Scio non esse nos Vergilios', I know that we are not Virgils. Still, the general atmosphere is one of cultural optimism: Rome is again a brilliant centre of art which need no longer yield to the Ancients. One cannot say that this was entirely self-delusion, since the Renaissance in Rome indeed produced most marvellous and lasting works of art—if not in literature, at least and most certainly in the visual arts and architecture.

Mgr Goritz used to receive his friends in his *vigna* on the slope of the Quirinal hill facing the Forum Romanum near the column of Trajan—in the words of Vida (397. 53–7, fo. HHiiii^{v}):

Traianique foro et Tarpeiae proximus arci
Excipit et vates et egregios heroas,
Quos mensis epulisque omnes dignatur opimis
Auricomas inter citros, quas ordine sevit
Ipse labore suo, putealique irrigat unda.

Near Trajan's Forum and the Tarpeian Rock he is host to both poets and distinguished men, inviting them all to a sumptuous banquet at his table, among the gold-laden orange trees, which he himself planted in order with his own hand and waters from the draw-well.

There he also exhibited his collection of antiquities, 'eruta Flaminiae, sunt eruta multa Latinae' (viz. 'Viae'),[10] as the Saxon guest Janus Hadelius puts it in one of his poems (368. 41, fo. Yiii^{r}). Both Blosius in his dedication and Silvanus in his fine poem *In Annales Corycianos* (398, fo. IIi^{r}–iiii^{r}) have left us charming descriptions of this garden, which no longer exists. It was situated on the Campo Carleo, a street of which only a few yards and a nameplate survive after the excavations of the imperial Fora.[11]

[10] Many of them unearthed near the Via Flaminia and the Via Latina.

[11] It is a pity that more recent historical sites so often have to disappear in favour of older

Let us enter the Vinea Coryciana and spend some moments in the merry company of poets, artists, scholars, and various servants of the Curia. Its relaxed atmosphere cannot better be described than in the words of a privileged visitor, the mysterious and unidentifiable German poet C. Silvanus Germanicus. This man celebrated the feast of Saint Anne in 1522 or 1523[12] in his lively and picturesque poem *In Annales Corycianos* (398, fos. II–IIiiii^{r}), the first part of which runs:

Anni curriculo perenniore
Annae iam rediit dies dicatus,
Quo Phoebea cohors, pii poetae
Diversum vario calore carmen
Edunt et populo exhibent legendum,
Qua divae Corytus locavit aram,
Atque arae statuas politiores
Eventu imposuit secundiore.
  Sed iam rite deae sacris peractis,
Traiani spacium Fori petamus,
Postes hic ubi Corytus recludit
Ornatos hedera sequaciore,
Et myrto Idalios olente odores,
Intertexta humilis coma genistae,
Et lauru pariter virentiore,
Invitatque viros disertiores
ut sumant patulis dapes sub umbris.
  Felix turba coit peritiorum,
Et docta serit allocutione
Sermones lepidos et elegantes.
Longa stant serie cibis onustae
Mensae, dumque sitim aridam Lyaeo,
Sed victo prius Albulae liquore,
Compescunt, Cerere et famem coercent,
Se non inferiora quisque narrant.
  Illinc sacra cohors chori perennis
Ad lyram numeros canit suaves,

strata, even if these earlier ones are little more than dust and rubble. Archaeology repeatedly turns out to be a real disaster for the Middle Ages, the Renaissance, and the Baroque in an ancient city such as Rome. One has only to compare the aspect of the Palatine Hill now and as it was in the high Renaissance when the Orti Farnesiani covered most of its ruins. Since 1882 archaeological scholarship has certainly gained an enormous amount of technical and historical knowledge, but at the cost of an equally great loss of art and beauty. Sometimes one wonders who was right, the Renaissance architects or the modern archaeologist.

[12] The date can be deduced from the fact that the second part of the poem is an encomium of the reigning Pope Hadrian VI.

Quos nec fulmina nec dies futuri
Absument carie vetustiore.
Hic ultra ingenio globos Olympi
Progressus retegit sacros libellos,
Et causas aperit quid Auctor orbis
Humana voluit tegi figura.
Alter sidereos refert recursus,
Quo vertiginis obviat rotatu
Noctis luna decus tenebricosae
Phoebeis radiis, ut igne semper
Fraterno niteat; quibus laboret
Defectu spaciis, quibusve signis
Portendat pluviam aut serenitatem,
Aut ventos per inane saevientes.
Hinc pensatur humi tenor iacentis,
Quantis oppida terminis recedant,
Quantus circuitus soli salique.
Naturae abdita funditus revellunt,
De rerum ordine disserunt vicissim
Docte, Iuppiter! et simul diserte.
Hospes interea hinc et inde felix
Mensas Corycius frequentat omnes,
Et large pateris manu Lyaeum
Propinat, lepidos ciens cachinnos.
Sol, qui praecipites agis quadrigas
In mundi medio, impetusque coeli
Adversos superas potente nisu,
Omnem et signiferum annuo meatu
Donec circuis, igneosque luci
Motu praerapido adiicis calores,
Omnes inde fluunt dies et anni,
Et rerum pariter perennis ordo,
Dic siquando parem virum coronam
Vidisti, aut epulas celebriores.

With the year's unfailing cycle, St Anne's day has now returned, the day on which Apollo's company, dutiful poets, variously inspired, produce divers poems and present them to all to read, at the spot where Goritz has erected an altar, felicitously placing on it the finest images.

Now that mass for the saint has duly been celebrated, let us make our way to Trajan's Forum where Goritz has thrown open his doors adorned with clinging ivy and with myrtle as fragrant as the Idalian sort, intertwined with foliage of the humble broom as well as with verdant bay. He invites these eloquent men to feast beneath the wide shade.

Happily the learned company assemble and in polished exchange begin their delightful and cultured discussions. The tables, covered with food, stand in extended row. Burning thirst they quench with wine already mixed with Tiber's liquor, and hunger they appease with Ceres' gifts. They keep up a conversation worthy of their distinction.

Over there the sacred company of the immortal choir is singing sweet songs to the lyre, which neither the thunderbolt nor days to come will ever consume in withering decay.

Here is one who, journeying in thought beyond the heavenly spheres, explains the Holy Scriptures and reveals the reasons why the Creator of the earth chose to put on human shape.

Another is discoursing on the orbits of the heavenly bodies: by what revolving rotation the moon, that ornament of the dark night, can meet the sun's rays and so beam continually with fraternal light; he explains the moon's various phases and by what signs it presages rain or fair weather or winds raging in the air. Next the extent of the level earth is examined, the distances between the cities, the circumference of continent and ocean.

They lay completely bare the secrets of nature and speak by turns about the order of things—good Lord, how learnedly and eloquently!

Meanwhile our host Goritz, in high spirits makes the rounds of every table, pouring wine into glasses in generous measure, putting all in a relaxed and laughing good mood.

O Sun, who drive your headlong four-horse chariot through the midst of heaven, and again ascend with powerful course your orbit, to pass through the whole Zodiac in your annual journey, bringing light and warmth with your swift movement, from which spring all days and years and the eternal order of things; tell us, did you ever see such a company of men or a more splendid banquet?

Poets singing, scholars discussing theological and philosophical issues, all of them with a glass of wine in one hand and delicious snacks in the other; a beaming host who fills the glasses, tells jokes and takes an immense delight in his gathering, and all that in the beatific shadow of fragrant green shrubs: who would not be happy to be a member of such an academic salon?

When Silvanus wrote this poem he certainly had in mind Statius, *Silvae* 1. 6, *Kalendae Decembres* (*On the first of December*), which celebrates a feast offered by the Roman emperor to the people. Both poems are written in the same metre, lightly running hendecasyllables, and both evoke a carefree merry gathering with plenty of wine and other good things. Yet Silvanus did not copy Statius and there are hardly any verbal reminiscences; poetry and learned conversation here take the place of simpler popular amusement—girls, dancing, and the like—there. At one point the German poet was thinking also of Horace. His address to the Sun (ll. 52 sqq.) is clearly one of the numerous offspring of the famous 'Alme Sol, curru nitido diem qui promis et celas' ('O bountiful Sun, who bring out and hide the bright daylight') of Horace's *Carmen saeculare*.

The unobtrusive way in which Silvanus used classical models and predecessors may well be typical for most of the Corycian poets. Nobody will doubt that they knew their classics very well, and there are indeed a number of obvious reminiscences of Catullus and Lucretius, Virgil and Horace, and other Roman classics. But those poets also knew that there is quite a difference between *imitatio virilis* and *imitatio puerilis*, between drawing inspiration from and enhancing a cherished model on one hand and slavishly copying it on the other.

To assess the actual extent of deliberate imitation in the *Coryciana* is at present difficult, as there is no modern edition providing us with a reliable *apparatus fontium*. So we must work with our own limited means of identification. But before I try to do something in that direction I want to express a caveat. Except where universally known and popular texts are involved or a longer passage is clearly being paraphrased or otherwise adapted, one must not too rashly leap in to identify imitation on the basis of verbal similarity or even identity. One tends to forget too easily that the strict prosodic and metrical rules and the selected vocabulary with which Latin poets usually operate will almost inevitably generate the same expressions in the same places in the verse as soon as the same ideas and feelings are put into words. Words such as *Venus*, *opus*, *decus*, *humus*, *manus*, *mare*, *meus*, *tuus*, and *suus* will be found at the end of innumerable dactylic pentameters; *Cupido*, *puella*, *papilla*, *voluptas*, *sagitta*, *terra*, *pontus*, and *cura* will fill the end of the hexameter. To interchange them is simply impossible. Experience has convinced me that many alleged reminiscences of less familiar texts are completely illusory because they are mere coincidences.

I cite an example from my own experience. In the fifties and the sixties of this century Harry C. Schnur (C. Arrius Nurus) used to cross the Atlantic every summer on board a cargo ship. To while away the tedium of the voyage he wrote Latin poems without the help of any book. One of them, 'At Sea', was first published in his *Nugulae* (New York, 1957), p. 35, then in the Vatican journal *Latinitas* 6 (1958), p. 20, and again in his volume *Pegasus Tolutarius* (Oudenaarde, 1962), p. 63. When I was helping him to prepare that little book for the press I noticed by chance that the beginning of the poem was clearly Statian. Indeed, compare Schnur, ll. 1–2:

> Mira quies pelagi. Non ullis concita ventis
> Aequora fulgescunt; nullis agitata procellis . . .

and Statius, *Silvae* 2. 2. 26–7:

> Mira quies pelagi: ponunt hic lassa furorem
> Aequora et insani spirant clementius Austri.[13]

When I asked him why he had not acknowledged this source as he had with other imitations he was absolutely struck dumb: he had never read Statius' *Silvae* in his whole life, a not very surprising fact. Indeed, how many professional classicists today can say they have read the *Silvae* entirely and thoroughly? And Schnur was a lawyer who had begun to study classical philology only at the age of 45 (in 1952).

Such experiences, of course, make one suspicious of the authenticity of many so-called reminiscences and dependences. In the following pages, therefore, I shall try to limit myself strictly to imitations of very popular classical models (passages which everybody had to study at school) and texts more than half a verse long. Even then one must remain cautious and ask oneself every time whether a conscious imitation is likely or whether a formula, once read in a text, has emerged more or less spontaneously in an identical metrical situation. Reminiscences in the *Coryciana* vary from similar metrical and/or verbal patterns to real quotations and close paraphrases—for all of which there are precedents in the works of the classical masters themselves.

Among similar patterns are verse endings of the type 'Haec . . . / . . . incensus meditabar amore' (374. 49, fo. Ziiii^{v}), where Lilius Gregorius Gyraldus may or may not have had in mind the Virgilian '. . . Dido meditatur amorem' (*Aeneid* 4. 171), and Baldassare Castiglione's '. . . laudes aequare canendo' (276. 10, fo. Riii^{v}), which of course resembles '. . . Phoebum superare canendo' in Virgil's famous *Eclogue* on Daphnis (5. 9). There can be little doubt that both Gyraldus and Castiglione were thoroughly familiar with, and perhaps knew by heart, those celebrated Virgilian texts. Still, it is less probable that they wanted us to find more in their own verses than a general Virgilian flavour. This seems to me to be true even where one can speak of direct citations, as in

[13] See also Lucan, *Bellum civile* 5.442–4: 'Saeva quies pelagi maestoque ignava profundo / stagna iacentis aquae, veluti deserta rigente / Aequora natura cessant . . .'.

line 3 of Hieronymus Corsinus' hymn to St Anne (poem 380, fos. BBiv^{v}–CC), 'Confluitis, multasque preces et dona ferentes', which partially repeats a Virgilian verse which has become an adage, 'Timeo Danaos, et dona ferentes'. There is no relationship of content between the two texts.

In the same poem, however, there is a passage where Tibullus really functions as an ancient model in a parallel situation. When Corsinus describes the feast of St Anne he uses a distich from Tibullus' *Elegy* 2. 1, a poem about a village festival. The Roman poet introduces the rustic holiday in the following words:

Luce sacra requiescat humus, requiescat arator
  Et grave suspenso vomere cesset opus.

On this holy day let the earth rest, let the ploughman rest,
Let the ploughshare be hung up and hard labour given over.

All the ingredients of the passage (the holy day, the rest of earth and ploughman, the image of the ploughshare, the giving-up of heavy toil) recur in Corsinus almost exactly in the same words and, as far as the different metre allows, in the same order:

. . . requiescat humus, requiescat arator
Vomere suspenso sacraque hac luce labores
Sint procul . . .

. . . Let the earth rest, let the ploughman rest,
Let the ploughshare be hung up and on this holy day let hard work
Be far from us . . .

Metrical influence is clear in the main change of words: *opus/labores*. The first neatly fits the end of the pentameter, *labores* the end of the hexameter, where neither *opus* nor *opera* would do. The choice of the word *labores* made the adjective *grave* (required by *opus*, which does not imply exertions) superfluous and it has accordingly been dropped.

There are more such intended reminiscences. Everybody knows the fourth *Eclogue* of Virgil, in Christian circles often interpreted as a prophecy of the birth of Jesus Christ. The child is addressed in line 49 as follows:

Cara deum soboles, magnum Iovis incrementum

Beloved offspring of the gods, great progeny of Jove

and this verse we immediately recognize in Bembo's Christianized address to Sansovino's child Jesus (36. 7, fo. Fi^{r}):

Aethereique puer magnum Patris incrementum

O child, great progeny of the heavenly Father

Only two words are identical, but the second, used for a child, is so rare and characteristic and its effect on the metre so striking that coincidence must here be ruled out.

Sometimes—but not very often—one comes across actual citations. A poet called Savoia (poem 43, fo. Fii^{v}–iii^{r}) wanted to express his incapacity to celebrate adequately the merits of the Corycian circle. Although writing in hendecasyllables, he managed to quote a verse which Virgil himself had already used twice (*Georgics* 2. 43–4 and *Aeneid* 6. 625–6):

Non mihi si linguae centum sint oraque centum
Ferrea vox . . .

Not if I had a hundred tongues and a hundred mouths
And a voice of iron . . .

With typical rhetorical *amplificatio* Savoia added another pair of conditional clauses (ll. 12–15):

Nam si sint mihi linguae et ora centum,
Si vox ferrea, et hinc sonem Maronem,
Hinc alto Cicerone contumescam,
Non laudem satis hos bonos poetas.

For if I had a hundred tongues and mouths, if I had a voice of iron, and now sounded like Virgil, now swelled like lofty Cicero, I could not do honour enough to these fine poets.

As usual, this rhetorical amplification of a famous verse is not very impressive. True genius is required for success with such word-plays.

A little unexpected is the conspicuous presence of Lucretius in a series of hymns to St Anne by Janus Vitalis (poem 375), Franciscus Centelles (377), Leonardus Tosta (378), Hieronymus Delius (379), and others. In those hymns the prooemium of the *De natura rerum* and the figure of Epicurus loom large in the words addressed to the Christian saint. I quote the first ten lines of Centelles to illustrate the point:

Anna, tuas laudes natamque altumque nepotem,
Regnatorem aulae aetheriae terraeque marisque,
Est animus libare; mihi tu, diva, canenti
Adsis et dubios audentis dirige gressus.
Te mortale genus, supera quod vescitur aura,

Te, dea, concelebrat, hominum per te genus omne
Iam sperare audet coeli aurea templa reclusi.
Tu natam das aetheriis quae regnat in astris
Quae gentes Herebique metu caecisque tenebris
Eripit et coelum reserat mortalibus aegris.

Anna, to sing your praises, your daughter and your exalted grandson, the king of heaven, earth, and sea, is my purpose. Tender me your help, O holy one, and lend your hand to the unsteady steps of my bold enterprise. This mortal race, which lives on earth, celebrates you, O holy one; through you all humanity dares now to hope that the golden temple of heaven will be opened. You give us a daughter who reigns among the stars of heaven, who frees the people from the fear of Hell and its black night, and unlocks the gates of heaven for wretched mortals.

Although this passage seems somewhat Virgilian in style,[14] at the same time it reads in places like a refashioning of Lucretius' hymn to Venus (1. 2–6):

Alma Venus . . .
Quae mare . . .
Concelebras, per te quoniam genus omne animantum
Concipitur . . .
Te, dea, te fugiunt . . .

Line 9, although clearly Christian in intention, recalls equally clearly Lucretius' Epicurean ideal concerning religion. Moreover, *coeli templa* is a recurrent Lucretian expression, sometimes used with an adjective, as in *coeli lucida templa* (1. 1014; 2. 1039). That adjective, however, had to be changed in Centelles' verse because otherwise it would have made one syllable too many in his hexameter. *Caecae tenebrae* also is a favoured word group of Lucretius (2. 55–6, 746, 798, and so on). Virgil prefers *caeca caligo*. Lucretius is also present when Janus Vitalis says to St Anne (ll. 28–9, fo. AA):

Ex tenebris tantis tam clarum educere lumen
Quae potuisti olim . . .

You who were able to bring forth so bright a light from such darkness . . .

Here he simply applies to her the praise of Epicurus with which Lucretius opens his third book:

E tenebris tantis tam clarum extollere lumen
Qui primus potuisti . . .

You who out of such darkness were first to lift up such a shining light . . .

[14] See e.g. verse endings such as '. . . si vescitur aura' (*Aeneid* 1. 546) and '. . . mortalibus aegris' (2. 268).

This is all the more remarkable because the reading of Lucretius was expressly forbidden under Leo X at the Synod of Florence in 1517.[15]

Apart from Virgil and Lucretius, and some explicit citations of other poets such as Martial and Juvenal in the prose dedication of the volume, Catullus and Horace are most prominently present.

Philippus Beroaldus the Younger offered a short ode (three stanzas) to Mgr Goritz (poem 21, fo. Diiii^r). The strophic Asclepiadic system clearly stems from Horace and reminds us of such famous poems as the odes to Pyrrha (1. 5) and other girls, the praise of the Bandusian spring (3. 13), or the hymn to Diana (1. 21). Yet Beroaldus' poem hardly makes an Horatian impression, and the only clear textual reminiscence is to Catullus. This is the poem:

> Vobis Corycius, maxima numina,
> Has ponit statuas, non sibi flagitans
> Ampli arbitria regni,
> Maiorem aut titulum ambiens.
>
> Nam quae vestra homini sancta dedit manus
> Sat laetum faciunt. Vos rogat ut diu
> His quae possidet uti
> Et vita incolumi queat.
>
> Audi, sacra cohors coelitum, et accipe
> Quas fert Corycius suppliciter preces,
> Si aevum puriter egit,
> Et si vos coluit pie.

To you, great powers, Goritz erects these images: he does not ask for himself the authority of exalted power, nor does he crave a more eminent title.

He is, indeed, content enough with what your holy hand has given him. He begs you that he may long enjoy it in a life free of stain.

Hear, you holy citizens of heaven, and receive the prayers that Goritz humbly offers you—provided that he has lived purely, and piously worshipped you.

The end of this ode is purely Catullan. Catullus' short elegy 76 is a prayer to the gods, exactly as Beroaldus' ode addresses Christ and the two saints. But whereas Catullus implores the gods to free him from an unpleasant disease, Beroaldus makes Corycius ask to be allowed to maintain his present happiness. In both cases, however, the arguments used as to why the gods or, for that

[15] See my article (n. 1 above), 41.

matter, the saints should hear the prayer are exactly the same, a pure and pious life. 'Me miserum aspicite et, si vitam puriter egi, / Eripite hanc pestem' ('Look upon my misery and if I have lived purely, remove this plague'), Catullus beseeches them (ll. 19–20), and Beroaldus echoes him in line 11: 'Si aevum puriter egi'. The Pherecratean verse is just one syllable shorter than the second hemistich of Catullus' hexameter; hence, instead of *vita*, its poetic equivalent *aevum*, which makes the elision of *si* possible. Catullus' last word is a reference to his piety (l. 26): 'O di, reddite mi hoc pro pietate mea!' ('O gods, give this to me on account of my piety!') Beroaldus uses the same argument, but the noun *pietas* is changed into a second *si* clause.

More Horatian, and yet wholly un-Horatian, is Poem 189 (fo. N–Ni^{v}) by Janus Vitalis. It consists of four stanzas made up of Horatian elements but quite unlike Horace himself. It has, to begin with, five-line stanzas, where Horace invariably sticks to four lines. The stanza is composed of four minor Asclepiadic verses followed by one Glyconic. Horace repeatedly uses three Asclepiadics to which he adds a Glyconic. Finally, Vitalis' Glyconic is a refrain, a device Horace never uses, but of which a classical example and justification could be found in Virgil's eighth *Eclogue*. Vitalis' poem reads as follows:

Qui per Carpàthii marmoris impetum et
Angustos scopulos, Acroceraunia
Iactatus, rapidum et per Siculum mare
Certus naufragio protinus arripi
  In saxa obvia Syrtibus.

Ad vos confugio, lucida sidera,
Iesu, Anna et Maria, o numina Coryti!
Terrae nam latuere, et mare turgidum
Huc illuc agitat subdita fragmina
  In saxa obvia Syrtibus.

Iam nox ignibus et fulminibus micat,
Dissolvit rabiem Sirius in ratem,
Eurus turgidus et vis Aquilonia
Adversis rapiunt carbasa flàtibus
  In saxa obvia Syrtibus.

Ad vos confugio, lucida sidera,
Ne mersus iaceam vortice languidus,
Illisus Boreae turbine et imbribus
Et quae concutiunt fulminibus ratem
  In saxa obvia Syrtibus.

I, who through the roughness of the Carpathian waves, the narrow rocks, the promontory of the Heights of Thunder and over the tempestuous Sicilian Sea am tossed, ever on the point of shipwreck on the cliffs and the sandbanks close by—

I seek refuge in you, bright stars, Jesus, Anna, Mary, Goritz's saving powers! There is no land in sight, and the swollen sea tosses the shattered fragments on to the cliffs and the sandbanks close by.

Lightning and thunderbolts flash through the night; Sirius has unleashed its ravening madness against the ship. A raging wind from the East and violence from the North sweep away in gusts of storm the sails on to the cliffs and the sandbanks close by.

I seek refuge in you, bright stars, from death by drowning, swallowed up by the waves, dashed by the Northern whirlwind and the pouring rain and the thunderbolts which hurl the boat upon the cliffs and the sandbanks close by.

Any reader familiar with Horace will immediately detect Horatian traces in this vow. Everyone knows *Ode* 1. 3, Horace's prayer to Venus and the Dioscuri that they might carry Virgil safely across the sea. That poem is written in alternating Glyconic and minor Asclepiadic verses; so the danger of excessive imitation was very real. The decision to create a new stanza with the same kind of verses reduced that danger and, in fact, only a few verbal borrowings subsist. The most obvious is line 2, which repeats, with a less fortunate variation of the adjective, Horace's line 20: 'infamis scopulos Acroceraunia'; and there is a trivial borrowing at the end of line 8 from Horace's 'qui vidit mare turgidum' (l. 19).

But the *lucida sidera*, taken from line 2 of Horace:

> Sic te diva potens Cypri,
> Sic fratres Helenae, lucida sidera . . .

May the mighty Cyprian goddess, may Helen's brothers, the bright stars . . .

are significant of the way in which ancient materials are reused for Christian purposes and filled with a new meaning: Venus and the Dioscuri are converted into Jesus and two women, St Anne and the Virgin Mary. Is there any fundamental difference, one must ask, between this literary Christianization of a pagan prayer and the reuse of the Pantheon as a church of the Blessed Virgin or, for that matter, the Roman Senate House as S. Adriano?

The comparison with Horace could be carried further or, rather, the differences could even be more exactly stated. Horace is praying for a friend who is undertaking a sea voyage and then goes on to meditate on the general rashness of men and their impiety. Vitalis prays for himself, although it is not entirely clear whether he is speaking of a real voyage from or to his native Palermo or of moral troubles. Whatever it was, the poem does not assume the universal dimension of Horace's ode, but remains within the boundaries of traditional Christian piety (as does one of Hutten's epigrams, which asks the same Corycian triad to heal his aching feet: poem 354, fo. Xiii^{v}).

A further accumulation of examples could easily run the risk of becoming as tedious as a protracted perusal of too many Corycian poems on one occasion. From what I have said I hope it will be clear that ancient Rome and its literature were very much alive in the hearts, the minds and the imagination of Medicean Rome. The classical city did not then so much represent a romantic, faraway past, as it does in some more recent novels of the *Quo vadis* type or in Pascoli's Latin poetry of about 1900. It was a very concrete counterpart of the present. The fact that the first Rome was pagan and the second Christian was not a real problem at that time. Nor must the use of ancient sources be interpreted—in my opinion—as a proof of the paganism of Medicean Rome. In the opinion of those Renaissance men pagan words, symbols, and stories could be blended harmoniously or filled with new Christian content without loss of Christian values. That such an acculturation was often not understood by Northerners visiting Rome (and by some modern scholars) is quite another story. Still, other Northerners eagerly joined their Roman and Italian brothers: Mgr Goritz himself was German, as were one of the editors of the *Coryciana*, C. Silvanus Germanicus, and a fair number of the contributors.

In order to see how old and new could function together in a way that may seem startling to us it is enough to read Laomedon Tardolus' hymn to St Anne (poem 381, fos. CC–CCii^{v}). This is, for the greater part, a prayer of Mgr Goritz to Anne, Mary, and Jesus. The prelate himself is introduced as a new Aeneas. In fact, line 4, 'Corycius geminas tollens ad sidera palmas' ('Goritz raising both his hands to heaven'), repeated *Aeneid* 1. 93, '[Aeneas] duplices tendens ad sidera palmas'; and the rest of the text also has a Virgilian colour. The wording is entirely classical and at times pagan, Hell being a *Tartareus lacus*. Yet the contents are definitely Christian. Goritz does not ask for riches, but for peace and above all (ll. 49–50):

hinc alacres post saecula longa caducis
Exuti membris, aeterna luce fruamur.

... May we then after long life, freed from our mortal limbs, enjoy eternal light.

Again, the Christian character of this prayer does not prevent the poet from concluding his hymn with a perfectly pagan image of Jupiter granting the requested favours:

Vix bene finierat, subito cum laevus ad auras
Cum tonitru fulgor supera demissus ab arce
Dat signum admissasque preces certo omine firmat.

He had hardly finished, when suddenly from the left a lightning-flash came down from heaven, accompanied by thunder: God had given a sign, and confirmed with a sure omen that their prayers had been granted.

At this point the stern theologian will knit his brows (and many have done so). That is to forget that the Corycians are, as one of them says explicitly, poets and not theologians. It is also to forget that the very first Christian Latin poet, Juvencus (early fourth century) also called God *Tonans* and heaven *Olympus* in his metrical paraphrase of the Gospels.

The challenge of ancient Rome, its poets and its monuments was a stimulant to the culture of Medicean Rome. Notwithstanding the far from brilliant political situation of those years, of which many an echo rings through in the *Coryciana*, many humanists thought that in art at least they were the equals of the glorious past or, to express it finally in Silvanus' words (poem 388, fo. EEiiii^{v}):

Nam si cuncta probe excuties, aequamus avorum
Ingenium artemque simul; documenta videre
Qui cupit, ad Coryti sacram se conferat aram.

Indeed, if you consider all correctly, we are the equals of our ancestors in talent and art alike. If you seek proof of this, come to Goritz's holy altar.

A hue of optimism pervades most of the Corycian poetry. For us, who know the doom that was soon to come and to smash that joyous world of poets and artists, the book leaves a feeling of sadness and sorrow. Not even the prayers of so many poets could save the good old Corycian gardener from the most bitter and least deserved of ends.

Index Locorum

2. BIBLICAL

General Index